Design for
the Real World

Victor Papanek

DESIGN FOR THE REAL WORLD

HUMAN ECOLOGY
AND SOCIAL CHANGE

With an introduction by
R. Buckminster Fuller

PANTHEON BOOKS
A DIVISION OF RANDOM HOUSE, NEW YORK

Acknowledgment is gratefully extended to the following for permission to reprint short excerpts from
their works:

Harold Ober Associates Incorporated: From Prologue
to *Prescription for Rebellion,* by Robert M. Lindner.
Copyright 1952 by Robert M. Lindner.

Penguin Books Ltd.: From "Lies" in *Selected Poems:
Yevtushenko,* translated by Robin Milner-Gulland
and Peter Levi, S.J. Copyright © 1962 by Robin
Milner-Gulland and Peter Levi.

The Museum of Modern Art: From *The Machine as
Seen at the End of the Mechanical Age,* by K. G.
Pontus Hultén, 1968.

FIRST AMERICAN EDITION

Printed by Halliday Lithograph Corp., West Hanover, Massachusetts. Bound by H. Wolff, New York

This volume is dedicated to my students,
for what they have taught me.

Introduction

THERE ARE wonderful friendships which endure both despite and because of the fact that the individuals differ greatly in their experiential viewpoints while each admires the integrity which motivates the other. Such friendships often are built on mutual reaction to the same social inequities and inefficiencies. However, having widely differing backgrounds, they often differ in their spontaneously conceived problem-solution strategies.

Victor Papanek and I are two such independently articulating friends who are non-competitive and vigorously cooperative. Long a Professor of Design at Purdue University, Victor Papanek now teaches that subject at the California Institute of the Arts. I am a "University Professor" at Southern Illinois University. I am a deliberate comprehensivist and do not operate in a department. Though I am a professor, I don't profess anything. The name of my professorship is "Comprehensive, Anticipatory Design-Science Exploration." I search for metaphysical laws governing both nature's a priori physical designing and the elective design initiatives of humans. It is typical of our friendship that I am permitted to write this introduction.

In this book, Victor Papanek speaks about everything as design. I agree with that and will elaborate on it in my own way.

To me the word "design" can mean either a weightless, metaphysical conception or a physical pattern. I tend to differentiate between design as a subjective experience, i.e.,

designs which affect me and produce involuntary and often
subconscious reactions, in contradistinction to the designs
that I undertake objectively in response to stimuli. What I
elect to do consciously is objective design. When we say
there is a design, it indicates that an intellect has organized
events into discrete and conceptual inter-patternings. Snow-
flakes are design, crystals are design, music is design, and
the electromagnetic spectrum of which the rainbow colors
are but one millionth of its range is design; planets, stars,
galaxies, and their contained behaviors such as the periodical regularities of the chemical elements are all design-ac-
complishments. If a DNA-RNA genetic code programs the
design of roses, elephants, and bees, we will have to ask
what intellect designed the DNA-RNA code as well as the
atoms and molecules which implement the coded programs.

The opposite of design is chaos. Design is intelligent or
intelligible. Most of the design subjectively experienced by
humans is a priori the design of sea waves, winds, birds,
animals, grasses, flowers, rocks, mosquitoes, spiders,
salmon, crabs, and flying fish. Humans are confronted with
an a priori, comprehensive, designing intellect which for instance has designed the sustenance of life on the planet we
call Earth through the primary impoundment of Sun energy
on Earth by the photosynthetic functioning of vegetation,
during which process all the by-product gases given off by
the vegetation are designed to be the specific chemical gases
essential to sustaining all mammalian life on Earth, and
when these gases are consumed by the mammals, they in
turn are transformed, again by chemical combinings and
disassociations, to produce the by-product gases essential to
the regeneration of the vegetation, thus completing a totally
regenerative ecological design cycle.

If one realizes that the universe is sum-totally an evolutionary design integrity, then one may be prone to acknowledge that an a priori intellect of infinitely vast con-
siderateness and competence is everywhere and everywhen
overwhelmingly manifest.

In view of a number of discoveries such as the ecological
regeneration manifest in the mammalian-vegetation inter-

exchange of gases, we can comprehend why responsibly thinking humans have time and again throughout the ages come to acknowledge a supra-human omniscience and omnipotence.

The self-regenerative scenario universe is an a priori design integrity. The universe is everywhere, and continually, manifesting an intellectual integrity which inherently comprehends all macro-micro event patterning and how to employ that information objectively with omni-consideration of all inter-effects and reactions. The universe manifests an extraordinary aggregate of generalized principles, none of which contradict one another and all of which are inter-accommodative, with some of the inter-accommodations exhibiting high exponential levels of synergetic surprise. Some of them involve fourth-power geometrical levels of energy interactions.

In addition to being a sailor, I am a mechanic. I carry a Journeyman's card in the International Association of Machinists and Aerospace Workers. I know how to operate all kinds of machine tools. I can always take a job in a machine shop, on a lathe, or work in a sheet metal shop on a stretch-press, press brake, et al. But I also have a wide experience in mass production tooling and in the general economics of the mass production concept. I understand the principle of tools that make tools.

As a youth on an island in Maine, I started designing spontaneously. I didn't draw something on paper and ask a carpenter to build it for me. I executed my own designing. I often had to make my own tools and procure my materials directly from the landscape. I would go into the woods and cut my own trees, dress them out, cure them, and then fashion them into their use form. I'm experienced in going from original conceptions, i.e., inventions—*ergo* unknown to others—to altering the environment in a complex of ways which are omni-considerate of all side effects on the altered environment. I am accustomed to starting from primitive conditions, where as far as one can see no other man has explored. I have learned how to rearrange the environment in such a way that it does various things for our society that we could not do before, such as building a dam which in turn produces a pond. I have been around for a half century to check on the adequacy of my earlier ecological considerations and their subsequent environmental inter-effects. No deleterious results are in evidence.

Because of that kind of experience, I am preoccupied with complex technology and all its social, industrial, economical, ecological, and physiological involvements. Industry involves all kinds of metallic alloys and plastics. I am interested not only in the chemistries but also in the tools necessary to accommodate the electrochemistry and metallurgy.

How does one take the prime design initiative—coping with all directly or indirectly related factors, from beginning to end of the problem? One must acquire mathematical knowledge and facility. When I'm building an airplane, I must know how to calculate the strength of its parts and their synergetic interaction in general assemblies and how to design their dynamic and static loading tests and how thereby to verify the theoretical calculations. I must understand the Bernoulli principles and Poisson's Law and whatever other general laws may be relevant. I must be competently familiar with all the civil and economic involvements of the aircraft use and maintenance, etc.

In the aircraft industry, for many years, the lead design teams were staffed with M.A. or Ph.D. engineers. Well developed in theory, these top engineers designed the airplanes and their many detailed parts. To get a little idea of the relative complexity involved, we note that ordinary single-family dwellings have about 500 types of parts. In respect to each type of part, there are usually a large number of mass-reproduced replicas of any one prototype part—such, for instance, as thousands of replicas of one type of finishing nail or multi-thousands of one type of brick. Automobiles involve an average of 5,000 types of parts and airplanes often involve 25,000 or more types of parts.

The lead-design-team calculations of aeronautical engineers embrace the stress and service behaviors of the finally assembled interactions of all the sub-assemblies of those parts as well as of the parts themselves. In the production and assembly of their end-products in single-family dwellings, automobiles, and airplane productions, the average deviation of finally assembled dimensions from the originally specified dimensions of their designers are plus or minus a quarter of an inch in dwellings—plus or minus a one-thousandth of an inch in automobiles—plus or minus one ten-thousandth of an inch in airplanes.

World War II was history's first war in which superior air power was the turning factor. Airplane production was inaugurated at an unprecedented magnitude. When in 1942 the U.S.A. came to designing and manufacturing multi-thousands of airplanes, it often developed that the design team engineers knew nothing about production methods and materials, as for instance they didn't realize that the aluminum alloy they specified came in certain standard production sizes with which they were unfamiliar. Time and again, during World War II, empty freight cars by the thousands were routed to aircraft companies to take away the ill-informed design engineers' waste scrap. More than half of all the tonnage of aluminum delivered to U.S. aircraft plants during World War II was carried away from those plants as scrap. The design engineers specified cutting the heart out of a sheet of metal for their special product and throwing away the two-thirds remainder because their theoretical studies showed an unreliability of rolled sheet along the sheets' edges. As a consequence, a separate corps of production engineers with the same theoretical competence as the designing engineers, but also deeply conversant with production practices as well as with the evolutionary frontier of newly available production tools, had to completely rework the original airplane-part designs to obtain equal strength and optimum end-performance but also suitability to available tool produceability. Production engineering involves "reserving tool time to make tools." There's a long forward scheduling. It takes much experience in the field of production to lay out the plant flow and tool set-up.

Then the comprehensive production engineer must understand the work of the men who make tools. The toolmaking

constitutes an extraordinary phase of industrial evolution
whether in Detroit's auto production or anywhere in the aircraft production world. The toolmakers are the invisible,
almost magical, "seven dwarfs" of industrial mass production. When the production engineer finds that no standard
tool exists which can do such-and-such an essential job,
then a good toolmaker and a good production engineer say,
"Yes, sir, we must evolve a tool to do so and so," and so they
do—thousands after thousands of times, and thus humanity's degree of freedom increases, and its days of life
are multiplied.

What is an impact extrusion? It is a vase-like receptacle
with a belly at its bottom and a narrow neck. You can take
clay and ram it in, bit by bit, through the top and narrow
neck. Finally you fill the whole bottom and then the neck
above it. An impact extrusion takes aluminum, for instance,
and squeezes it into such a vase-like vessel while periodically impacting it violently to spread it outward at its base.
After forming, the vase-like vessel opens, in separate vertical parts, to free the completed and contained part. This is
a typical operation of production engineering. Production
engineers have to know how to heat-treat and anneal their
parts and whether various alloyed metals can stand rework-
ing, punching, drawing, and how far they can transform
before the material breaks apart (crystallizes) and whether
further heat-treating may make further transforming possible. Production engineering calls for an artist-scientist-in-
ventor with enormous experience.

When complex assemblies are finally produced, the comprehensive designer must know how to get them to wherever
they need to go. He may need to put his products in wooden
crates so they can be moved safely from "here to there," and
he has to know what the freight regulations are, etc.

In my early days in New York before World War I, there
were relatively few automobiles and no motor trucks. There
were a few very small electric vans. Trucking was done primarily by horses and drays. The men driving horses were
pretty good at their driving. They were usually illiterate and
often drunk. Bringing up his dray with or without helper,

the driver was interested only in getting his rig loaded or un-loaded. This didn't call for intellectual talent, and it didn't call for anybody interested in the product. The drivers and luggers were just moving this and that from here to there. I saw trucks being loaded and unloaded wham-bam; anything would do. The truckers and luggers didn't know what was inside the packages, so they just piled them here or there as their fancy pleased. Sometimes the loads would slide off the dray.

New York was full of little manufacturies that needed products delivered. I guess that 25 per cent of everything that went on any truck in those days was destined to ruin. There was an assumption on the part of the manufacturer and the people receiving the goods that 25 per cent of every-thing received would be destroyed in handling. The idea of packing products in corrugated cardboard cartons, as we do today, had not yet evolved. The then existent cardboard boxes were poorly designed and often broke open. In order to be sure that expensive things didn't get hurt, merchants put them in very heavy and expensive wooden crating. They tried to make their crating indestructible. If the load fell off a truck, they hoped it wasn't going to get hurt. The en-gineers who were shipping on ocean steamships designed special ocean steamship crating calculated to withstand handling in great nets elevated by enormous booms which swung the loads aboard and banged them down into the hold. Often these crates and loaded slings would crash against the ship's side. Insurance companies began to give rebates or premiums to clients who designed better cases. Thus the container business began to thrive. After World War II came the foam-formed packaging. Since then, tele-vision sets, cameras, et al. have been neatly packed in shock-proof plastic pre-forms.

During World War II, many enormous, partly completed airplane sub-assemblies had to be moved from here to there. These parts were very valuable. To make crates large and strong enough for these parts was formidably difficult. The comprehensive engineers built special trucks to do one single task. Inside these trucks, they installed special jigs that would securely hold a particular product. This was called jig-shipping. The uncrated products were bolted safely in place.

Out of the jig-shipping developed the design of standard railway boxcar or trailer truck containers which could contain jig-shipping fixtures and could be loaded interchangeably onto railway flatcars, onto trailer wheels, or onto specially designed ocean-going ships devoted exclusively to container shipping. Comprehensive Anticipatory Design Science embraces the foregoing design-evolution initiatives. It must be responsible all the way from the geographical points at which the raw resources occur in nature, and that means in remote places all around the world. Comprehensive Anticipatory Design Science must be responsible for designing every process all through the separating, mining, refining stages and their subsequent association into alloys, and subsequent forming into products. Comprehensive Anticipatory Design Science must learn how you go from ingots into rolled sheet and to convert the latter into the next form. The Comprehensive Anticipatory Design Scientist knows that forming steel into the intermediate merchandising forms of tubes, angles, I-beams, sheet and plate can often be avoided, if the metals are produced originally by their end-user.

World War I saw the beginning of industrially produced alloys. World War II employed an enormous variety of steel and aluminum alloys. To accommodate the multiplicity of design requirements, the steel and aluminum manufacturers formed angles, channels, I-beams, T's, Z's, in a vast assortment of sizes and alloys. This meant that the aircraft plant was stocked with full bins of all kinds of sections of different sizes and alloys of metals all color-coded. These alloys did not exist in World War I. There was mild steel and piano wire steel and a few others. World War II saw so many different kinds of steels and aluminum developed that were designated by decimal code numbers whose types ran into the thousands. All kinds of complex color-band codes were also used for special classes of material. The aircraft plant bins were full of a vast variety of rods, bars, angles, channels, hat-sections, et al. As parts were cut out of these stocks, there was waste. Because manufacturers of original metal stock had to have standard sizes, the aircraft producers had to cut big parts into little parts from one kind of section or another.

As a consequence of all that waste and duplicated effort, after World War II the aircraft industry production techniques began to change rapidly. Donald Douglas, founder and pioneer of the DC–3 and its "DC–" descendants, said, "I'm never again going to have a design engineer who isn't also a production engineer. We must eliminate these two stages." Another factor that induced method changes was that complex alloying began to increase even more rapidly after World War II. The computer made it possible to cope with more complex problems. Aerospace research brought about new knowledge which resulted in unprecedented advances in alloys. To produce the new jets or rockets, metals were needed with strength and heat-resistance capabilities not as yet known to exist. Therefore, metals had to be developed which could withstand the re-entry heats and structural stresses of rocket capsules returning into the atmosphere which generated fantastic degrees of heat as they rushed back into the air at thousands of miles per hour. Metals had to be strong enough to hold the capsule together. Thus a new industrial production era began in which the designing engineer said, "We're going to have to have such-and-such a capability metal which is not as yet known to exist." So for the first time in history, metallurgists aided by computers were able to produce enough knowledge regarding nature's fundamental associabilities and disassociabilities to be able to design new, unprecedented metals. That was strictly a post–World War II event.

Up to this moment research scientists had from time to time made discoveries of new alloys—their discoveries could not be predicted. At mid-twentieth century, 1950, humans began to design specific metals for specific functions to be produced in the exact amounts and formed instantly in the final use shape. Thus a truly new phase of comprehensive design began wherein a pre-specified, unprecedented metal was produced immediately in its ultimate use shape. The airspace production no longer had to go through the intermediary phases of finding the nearest type of special alloy, and the special "dimension" angle, or Z-bar, which must then go to the machine shop for special cutting and further forming. The swift evolutions in design strategy are not even taught in the engineering schools, for the airspace technology is often "classified," and the engineering school professors had no way of learning about the changes.

At M.I.T. there are buildings full of rooms, and rooms full of yesterday's top priority machinery that is now utterly obsolete. They have a vast graveyard of technology. The students don't want to take classes in mechanical engineering any more because they have heard that what they learn is going to be obsolete before they graduate. These evolutionary events cover all phases of technology and the physical sciences. Victor Papanek's book conducts a mass funeral service for a whole segment of now obsolete professionals.

These now swiftly accelerating events in the design and production competence of humanity with which the Comprehensive Anticipatory Design Scientist is concerned are symptomatic of far greater evolutionary transformations in the life of humanity aboard our planet.

We are transforming from a five-and-one-half-million-year period of humanity isolated in small tribes scattered so far apart as to have no knowledge of one another. At the tail-end of that period there developed a ten-thousand-year period in which humans built fortified citadels and a few fortified cities commanding scarce and rich farming areas— all of which were as yet so remote from one another that existence of other such city-states was only legendary hearsay to the dwellers within any one such city. Once in a decade or century, droughts, floods, fires, pestilence, and other disasters in one such city-state sent its inhabitants migrating away seeking new lands to support them. Great wars occurred as they discovered and invaded the cultivated lands of others. Halfway through that last ten thousand years of history, which opened with the city-states, the evolution from grass-and-pitch boats, inflated pigskin floats, log

rafts and dugouts developed into powerful, keeled and rib-
bed, deep-bellied vessels which attained high-seas-keeping
capability and with celestial navigation attained the competence to traverse the great oceans with cargoes vaster than
could be carried overland on the backs of animals. Water
covers three quarters of the Earth, whose three seemingly
separate oceans are only joined together, free of ice, around
the Antarctic continent many thousands of miles away from
the 95 per cent of humanity living in the northern lands of
our planet.

With this discovery that all the oceans were interconnected, there began the integratability of world-around resources whose alloyable associabilities generate ever higher
physical advantage for ever greater numbers of humanity
and give rise to the phenomenon industrialization. Industrialization is the integration of all the known history of experiences of all of humanity resolved into scientific principles which enable the doing of ever more comprehensively
adequate tasks with ever less investment of human time,
kilowatts of power, and pounds of material per each accomplished function, accomplished primarily from energies
other than those impounded by today's or yesterday's vegeta-
tion-capture of Sun energy.

The last five hundred years of humanity aboard our
planet have witnessed the at first gradual and now ever
swifter development of world-embracing industrialization.
Early regeneration of human life aboard our planet was sustained exclusively by the a priori vegetation, fish, and land
animal flesh. These foods were hunted, hand-picked, or
hand-cultivated. Then came irrigation, and after World War
I, mechanization of farming tools and vehicles such as the
plows, autos, and reapers which involved taking the fossil
fuels from nature's terrestrial storage battery, to start and
keep their engines moving. In the last one hundred years
electromagnetics and production steel have permitted man
to harness some of the limitless, eternally transforming
energy of the universe's main engines. What is going on is
analogous to using the storage battery to actuate our self-
starter to get us hooked up with the inexhaustible main en-

gine of the universe which will quite incidentally recharge the Earth's fossil-fuel storage battery. In the last fifty years we have started to establish a world-around integratable energy-distributing network which will soon be switched into the inexhaustible celestial energy system of the infinitely regenerative universe.

To fulfill his potential usefulness to humanity, the Comprehensive Anticipatory Design Scientist must multiply his numbers to permit the conversion of humanity from a you-or-me ignorance status to the omni-successful education and sustenance of all humanity. That omni-success has now become technically feasible, but is frustrated by humanity's clinging ignorantly to the inherently shortsighted one-year accounting system which was suitable only to yesterday's life support which was entirely dependent upon "this year's" perishable crop of Sun energy which was then exclusively impounded by land-borne vegetation and water-borne algae.

Now we have available the inexhaustible, gravitationally generated, tidal power of the world ocean to feed into our soon-to-be-accomplished world-around electromagnetic power network to be fed also by wind and direct Sun.

No longer is it valid to say, "We can't afford to spend," which concept was fundamentally generated by the truly expendable, because highly perishable, easily exhaustible, exclusively biological impoundment of Sun energy. Now we are throwing the switch to connect humanity into the universe's eternally self-regenerative system. This brings with it the ability to say we have attained unlimited ability to regenerate local life aboard our planet and within its ever expanding celestial neighborhood. Designing the new accounting system is the task of the Comprehensive Anticipatory Design Scientist. The new economic accounting system must make it eminently clear that whatever we need to do, that we know how to do, we can afford to do.

It is the growing pains of this epochal transition which give rise to the conditions with which Victor Papanek deals so effectively in this book. He is lowering the asbestos curtain on the historical scene of an Earth-bound humanity universally frustrated by the last days of omni-specialization. Omni-specialization by the educational system was yesterday's physical tyrants' means of effecting their omni-divide-and-conquer strategy. If humanity is to survive aboard our planet, it must become universally literate and preoccupied with inherently cooperative Comprehensive Anticipatory Design Science in which every human is concerned with accomplishing the comfortably sustainable well-faring of all other humans.

R. Buckminster Fuller
Carbondale, Illinois

Preface

THERE ARE PROFESSIONS more harmful than industrial design, but only a very few of them. And possibly
only one profession is phonier. Advertising design, in persuading people to buy things they don't need, with money
they don't have, in order to impress others who don't care, is
probably the phoniest field in existence today. Industrial
design, by concocting the tawdry idiocies hawked by ad-
vertisers, comes a close second. Never before in history have
grown men sat down and seriously designed electric hairbrushes, rhinestone-covered file boxes, and mink carpeting
for bathrooms, and then drawn up elaborate plans to make
and sell these gadgets to millions of people. Before (in the
"good old days"), if a person liked killing people, he had to
become a general, purchase a coal mine, or else study nu-
clear physics. Today, industrial design has put murder on a
mass production basis. By designing criminally unsafe automobiles that kill or maim nearly one million people around
the world each year, by creating whole new species of permanent garbage to clutter up the landscape, and by choosing
materials and processes that pollute the air we breathe, designers have become a dangerous breed. And the skills
needed in these activities are taught carefully to young
people.

In an age of mass production when everything must be planned and designed, design has become the most powerful tool with which man shapes his tools and environments (and, by extension, society and himself). This demands high social and moral responsibility from the designer. It also demands greater understanding of the people by those who practice design and more insight into the design process by the public. Not a single volume on the responsibility of the designer, no book on design that considers the public in this way, has ever been published anywhere.

In February of 1968, *Fortune* magazine published an article that foretold the end of the industrial design profession. Predictably, designers reacted with scorn and alarm. But I feel that the main arguments of the *Fortune* article are valid. It is about time that industrial design, *as we have come to know it*, should cease to exist. As long as design concerns itself with confecting trivial "toys for adults," killing machines with gleaming tailfins, and "sexed-up" shrouds for typewriters, toasters, telephones, and computers, it has lost all reason to exist.

Design must become an innovative, highly creative, cross-disciplinary tool responsive to the true needs of men. It must be more research-oriented, and we must stop defiling the earth itself with poorly-designed objects and structures.

For the last ten years or so, I have worked with designers and student design teams in many parts of the world. Whether on an island in Finland, in a village school in Indonesia, an air-conditioned office overlooking Tokyo, a small fishing village in Norway, or where I teach in the United States, I have tried to give a clear picture of what it means to design within a social context. But there is only so much one can say and do, and even in Marshall McLuhan's electronic era, sooner or later one must fall back on the printed word.

Included in the enormous amount of literature we have about design are hundreds of "how-to-do-it" books that address themselves exclusively to an audience of other designers or (with the gleam of textbook sales in the author's

eye) to students. The social context of design, as well as the
public and lay reader, is damned by omission.

Looking at the books on design in seven languages, covering the walls of my home, I realized that the one book I
wanted to read, the one book I most wanted to hand to my
fellow students and designers, was missing. Because our
society makes it crucial for designers to understand clearly
the social, economic, and political background of what they
do, my problem was not just one of personal frustration. So
I decided to write the kind of book that I'd like to read.

This book is written from the viewpoint that there is
something basically wrong with the whole concept of patents
and copyrights. If I design a toy that provides therapeutic
exercise for handicapped children, then I think it is unjust
to delay the release of the design by a year and a half, going
through a patent application. I feel that ideas are plentiful
and cheap, and it is wrong to make money off the needs of
others. I have been very lucky in persuading many of my
students to accept this view. Much of what you will find as
design examples throughout this book has never been
patented. In fact, quite the opposite strategy prevails: in
many cases students and I have made measured drawings
of, say, a play environment for blind children, written a
description of how to build it simply, and then mimeo-
graphed drawings and all. If any agency, anywhere, will
write in, my students will send them all the instructions
free of charge. I try to do the same myself. An actual case
history may explain this principle better.

Shortly after leaving school nearly two decades ago, I designed a coffee table based on entirely new concepts of
structure and assembly. I gave a photograph and drawings
of the table to the magazine *Sunset*, which printed it as a
do-it-yourself project in the February, 1953, issue. Almost
at once a Southern California furniture firm, Modern Color,
Inc., "ripped-off" the design and went into production. Admittedly they sold about eight thousand tables in 1953. But
now it is 1970. Modern Color has long since gone bankrupt, but *Sunset* recently reprinted the design in their book

"Transite Table," author's design, courtesy: *Sunset* magazine.

Furniture You Can Build, so people are still building the table for themselves.

Thomas Jefferson himself entertained grave doubts as to the philosophy inherent in a patent grant. At the time of his invention of the hemp-break, he took positive steps to prevent being granted a patent and wrote to a friend: "Something of this kind has been so long wanted by cultivators of hemp that as soon as I can speak of its effect with certainty, I shall probably describe it anonymously in the public papers in order to forestall the prevention of its use by some interloping patentee."

I hope this book will bring new thinking to the design process and start an intelligent dialogue between designer and consumer. It is organized into two parts, each six chapters long. The first part, "Like It Is," attempts to define and criticize design as it is practiced and taught today. The six chapters of "How It Could Be" address themselves to the late Dr. Robert Lindner, of Baltimore, with whom I give the reader at least *one* newer way of looking at things in each chapter.

I have received inspiration and help in many parts of the world, over many years, in forming the ideas and ideals that made the writing of this book so necessary. I have spent large chunks of time living among Navahos, Eskimos, and Balinese, as well as spending nearly one third of each of the last seven years in Finland and Sweden, and I feel that this has shaped my thoughts.

In Chapter Four, "Do-It-Yourself Murder," I am indebted to the late Dr. Robert Lindner of Baltimore, with whom I corresponded for a number of years, for his concept of the "Triad of Limitations." The idea of *kymmenykset* was first formulated by me during a design conference on the island of Suomenlinna in Finland in 1968. The word *Ujamaa,* as a simple way of saying "we work together and help each other without colonialism or neo-colonial exploitation," was supplied in Africa during my UNESCO work.

Mr. Harry M. Philo, an attorney from Detroit, is responsible for many of the examples of unsafe design cited in Chapter Five.

Much in Chapter Eleven, "The Neon Blackboard," reflects similar thinking by my two good friends, Bob Malone of Connecticut, and Bucky Fuller.

Four people are entitled to special thanks. Walter Muhonen of Costa Mesa, California, because the example set by his life has kept me going, even though my goals seemed unattainable. He taught me the real meaning of the Finnish word *sisu.* Patrick Decker of College Station, Texas, for persuading me to write this book. "Pelle" Olof Johansson of Halmstad and Stockholm, Sweden, for arguing the fine points of design with me, long into many nights; and for making the actual completion of this book's first Swedish edition possible. My wife, Harlanne, helped me to write what I wanted to say, instead of writing what seemed to sound good. Her searching questions, criticism, and encouragement often made all the difference.

The incisive thinking and the help of my editor, Verne Moberg, have made this revised edition sounder and more direct.

In an environment that is screwed up visually, physically, and chemically, the best and simplest thing that architects, industrial designers, planners, etc., could do for humanity would be *to stop working entirely*. In all pollution, designers are implicated at least partially. But in this book I take a more affirmative view: It seems to me that we can go beyond not working at all, and work positively. Design can and must become a way in which young people can participate in changing society.

Ever since the German Bauhaus first published its fourteen slender volumes around 1924, most books have merely repeated the methods evolved there or added frills to them. A philosophy more than half a century old is out of place in a field that must be as forward-looking as this.

As socially and morally involved designers, we must address ourselves to the needs of a world with its back to the wall while the hands on the clock point perpetually to one minute before twelve.

Helsinki—Singaradja (Bali)—Stockholm
1963–1971

Contents

PART

ONE

LIKE IT IS

1

WHAT IS DESIGN?

A Definition of Design and

the Function Complex

The wheel's hub holds thirty spokes
Utility depends on the hole through the hub.
The potter's clay forms a vessel
It is the space within that serves.
A house is built with solid walls
The nothingness of window and door alone
 renders it useable,
That which exists may be transformed
What is non-existent has boundless uses.

LAO-TSE

ALL MEN are designers. All that we do, almost all the
time, is design, for design is basic to all human activity. The
planning and patterning of any act towards a desired, foreseeable end constitutes the design process. Any attempt to
separate design, to make it a thing-by-itself, works counter
to the inherent value of design as the primary underlying
matrix of life. Design is composing an epic poem, executing
a mural, painting a masterpiece, writing a concerto. But design is also cleaning and reorganizing a desk drawer, pulling
an impacted tooth, baking an apple pie, choosing sides for
a back-lot baseball game, and educating a child.

Design is the conscious effort to impose meaningful order.

The order and delight we find in frost flowers on a
window pane, in the hexagonal perfection of a honeycomb,

in leaves, or in the architecture of a rose, reflect man's pre-occupation with pattern, the constant attempt to understand an ever-changing, highly complex existence by imposing order on it—but these things are not the product of design. They possess only the order we ascribe to them. The reason we enjoy these and other things in nature is that we see an economy of means, simplicity, elegance, and an essential rightness in them. But they are not design. Though they have pattern, order, and beauty, they lack conscious intention. If we call them design, we artificially ascribe our own values to an accidental side issue. The streamlining of a trout's body is aesthetically satisfying to us, but to the trout it is a by-product of swimming efficiency. The aesthetically satisfying spiral growth pattern found in sunflowers, pineapples, pine cones, or the arrangement of leaves on a stem can be explained by the Fibonacci sequence (each member is the sum of the two previous members: 1, 1, 2, 3, 5, 8, 13, 21, 34 . . .), but the plant is only concerned with improving photosynthesis by exposing a maximum of its surface. Similarly, the beauty we find in the tail of a peacock, although no doubt even more attractive to a peahen, is the result of intraspecific selection (which, in the case cited, may even ultimately prove fatal to the species).

Intent is also missing from the random order system of a pile of coins. If, however, we move the coins around and arrange them according to size and shape, we add the element of intent and produce some sort of symmetrical alignment. This symmetrical order system is a favorite of small children, unusually primitive peoples, and some of the insane, because it is so easy to understand. Further shifting of the coins will produce an infinite number of asymmetrical arrangements which require a higher level of sophistication and greater participation on the part of the viewer to be understood and appreciated. While the aesthetic values of the symmetrical and asymmetrical designs differ, both can give ready satisfaction since the underlying intent is clear. Only marginal patterns (those lying in the threshold area between symmetry and asymmetry) fail to make the designer's intent clear. The ambiguity of these "threshold cases" produces a feeling of unease in the viewer. But apart from these threshold cases there are an infinite number of possible satisfactory arrangements of the coins. Importantly, none of these is the one right answer, though some may seem better than others.

Shoving coins around on a board is a design act in miniature because design as a problem-solving activity can never, by definition, yield the one right answer: it will always produce an infinite number of answers, some "righter" and some "wronger." The "rightness" of any design solution will depend on the meaning with which we invest the arrangement.

Design must be meaningful. And "meaningful" replaces the semantically loaded noise of such expressions as "beautiful," "ugly," "cool," "cute," "disgusting," "realistic," "obscure," "abstract," and "nice," labels convenient to a bankrupt mind when confronted by Picasso's "Guernica," Frank Lloyd Wright's Fallingwater, Beethoven's *Eroica*, Stravinsky's *Le Sacre du printemps*, Joyce's *Finnegans Wake*. In all of these we respond to that which has meaning.

The mode of action by which a design fulfills its purpose is its function.

"Form follows function," Louis Sullivan's battle cry of the 1880's and 1890's, was followed by Frank Lloyd Wright's "Form and function are one." But semantically, all the statements from Horatio Greenough to the German Bauhaus are meaningless. The concept that what *works* well will of necessity *look* well, has been the lame excuse for all the sterile, operating-room-like furniture and implements of the twenties and thirties. A dining table of the period might have a top, well proportioned in glistening white marble, the legs carefully nurtured for maximum strength with minimum materials in gleaming stainless steel. And the first reaction on encountering such a table is to lie down on it and have your appendix extracted. Nothing about the table says: "Dine off me." *Le style international* and *die neue Sachlichkeit* have let us down rather badly in terms of human value.

Le Corbusier's house as *la machine à habiter* and the packing-crate houses evolved in the Dutch *De Stijl* movement reflect a perversion of aesthetics and utility.

"Should I design it to be functional," the students say, "or to be aesthetically pleasing?" This is the most heard, the most understandable, and the most mixed-up question in design today. "Do you want it to look good, or to work?" Barricades erected between what are really just two of the many aspects of function. It is all quite simple: aesthetic value is an inherent *part* of function. A simple diagram will show the dynamic actions and relationships that make up the function complex:

Figure 1: THE FUNCTION COMPLEX

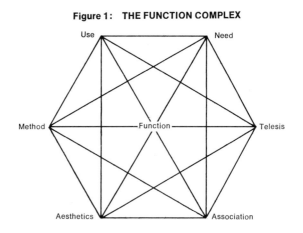

It is now possible to go through the six parts of the function complex (above) and to define every one of its aspects.
METHOD: The interaction of *tools, processes,* and *materials.* An honest use of materials, never making the material seem that which it is not, is good method. Materials and tools must be used optimally, never using one material where another can do the job less expensively and/or more efficiently. The steel beam in a house, painted a fake wood grain; the molded plastic bottle designed to look like expensive blown glass; the 1967 New England cobbler's bench reproduction

("worm holes $1 extra") dragged into a twentieth-century living room to provide dubious footing for martini glass and ash tray: these are all perversions of materials, tools, and processes. And this discipline of using a suitable method extends naturally to the field of the fine arts as well. Alexander Calder's "The Horse," a compelling sculpture at the Museum of Modern Art in New York, was shaped by the particular material in which it was conceived. Calder decided that boxwood would give him the specific color and texture he desired in his sculpture. But boxwood comes only in rather narrow planks of small sizes. (It is for this reason that it traditionally has been used in the making of small boxes: hence its name.) The only way he could make a fairsized piece of sculpture out of a wood that only comes in small pieces was to interlock them somewhat in the manner of a child's toy. "The Horse," then, is a piece of sculpture, the aesthetic of which was largely determined by method. For the final execution at the Museum of Modern Art Calder chose to use thin slats of walnut, a wood similar in texture.

When early Swedish settlers in what is now Delaware decided to build, they had at their disposal trees and axes. The *material* was a round tree trunk, the *tool* an ax, and the *process* a simple kerf cut into the log. The inevitable result of this combination of tools, materials, and process is a log cabin.

From the log cabin in the Delaware Valley of 1680 to Paolo Soleri's desert home in twentieth-century Arizona is no jump at all. Soleri's house is as much the inevitable result

Alexander Calder: "The Horse" (1928). Walnut, 15½ x 34¾. Collection The Museum of Modern Art, New York. Acquired through the Lillie P. Bliss Bequest.

Paolo Soleri: Carved earth form for the original drafting room and interior of the ceramics workshop. Photos by Stuart Weiner.

of tools, materials, and processes as is the log cabin. The peculiar viscosity of the desert sand where Soleri built his home made his unique method possible. Selecting a mound of desert sand, Soleri criss-crossed it with V-shaped channels cut into the sand, making a pattern somewhat like the ribs of a whale. Then he poured concrete in the channels, forming, when set, the roof-beams of the house-to-be. He added a concrete skin for the roof and bulldozed the sand out from underneath to create the living space itself. He then completed the structure by setting in car windows garnered from automobile junkyards. Soleri's creative use of tools, materials, and processes was a *tour de force* that gave us a radically new building method.

Dow Chemical's "self-generating" styrofoam dome is the product of another radical approach to building methods. The foundation of the building can be a 12-inch-high circular retaining wall. To this wall a 4-inch-wide strip of styrofoam is attached, which raises as it goes around the wall from zero to 4 inches in height, forming the base for the spiral dome. On the ground in the center, motorized equipment operates two spinning booms, one with an operator and the other holding a welding machine. The booms move around, somewhat like a compass drawing a circle, and they rise with a spiraling motion at about 30 feet a minute. Gradually they move in towards the center. A man sitting in the saddle feeds an "endless" 4 x 4-inch strip of styrofoam into the welding machine, which heat-welds it to the previously hand-laid styrofoam. As the feeding mechanism follows its circular, rising, but ever-diminishing diameter path, this spiral process creates the dome. Finally, a hole 36 inches in diameter is left in the top, through which man, mast, and moment arm can be removed. The hole is then closed with a clear plastic pop-in bubble or a vent. At this point the structure is translucent, soft, but still entirely without doors or windows. The doors and windows are then cut (with a minimum of effort; in fact the structure is still so soft that openings could be cut with one's fingernail), and the structure is sprayed inside and out with latex-modified concrete. The dome is ultra-lightweight, is secured to withstand high wind speeds and great snow loads, is vermin-proof, and inexpensive. Several of these 54-foot-diameter domes can be easily joined together into a cluster.

All these building methods demonstrate the elegance of solution possible with a creative interaction of tools, materials, and processes.

USE: "Does it work?" A vitamin bottle should dispense pills singly. An ink bottle should not tip over. A plastic-film package covering sliced pastrami should withstand boiling water. As in any reasonably conducted home, alarm clocks seldom travel through the air at speeds approaching five hundred miles per hour, "streamlining" clocks is out of place. Will a cigarette lighter designed like the tailfin of an automobile (the design of that automobile was copied from a pursuit plane of the Korean War) give more efficient service? Look at some hammers: they are all different in weight, material, and form. The sculptor's mallet is fully round, permitting constant rotation in the hand. The jeweler's chasing hammer is a precision instrument used for fine work on metal. The prospector's pick is delicately balanced to add to the swing of his arm when cracking rocks.

The ball-point pen with a fake polyethylene orchid surrounded by fake styrene carrot leaves sprouting out of its top, on the other hand, is a tawdry perversion of design for use.

But the results of the introduction of a new device are never predictable. In the case of the automobile, a fine irony developed. One of the earliest criticisms of the car was that, unlike "old Dobbin," it didn't have the sense "to find its way home" whenever its owner was incapacitated by an evening of genteel drinking. No one foresaw that mass acceptance of the car would put the American bedroom on wheels, offering everyone a new place to copulate (and privacy from supervision by parents and spouses). Nobody expected the car to accelerate our mobility, thereby creating the exurban sprawl and the dormitory suburbs that strangle our larger cities; or to sanction the killing of fifty thousand people per annum, brutalizing us and making it possible, as Philip Wylie says, "to see babies with their jaws ripped off on the corner of Maine and Maple"; or to dislocate our societal groupings, thus contributing to our alienation; and to put every yut, yahoo, and prickamouse from sixteen to sixty in permanent hock to the tune of $80 a month. In the middle forties, no one foresaw that, with the primary use function of the automobile solved, it would emerge as a combination status symbol and disposable, chrome-plated codpiece. But two greater ironies were to follow. In the early sixties, when people began to fly more, and to rent standard cars at their destination, the businessman's clients no longer saw the car he owned and therefore could not judge his "style of life" by it. Most of Detroit's baroque exuberance subsided, and the automobile again came closer to being a transportation device. Money earmarked for status demonstration was now spent on boats, color television sets, and other ephemera. The last irony is still to come: with carbon monoxide fumes poisoning our atmosphere, the electric car, driven at low speeds and with a cruising range of less than one hundred miles, reminiscent of the turn of the century, may soon make an anachronistic comeback. Anachronistic because the days of individual transportation devices are numbered.

The automobile gives us a typical case history of seventy years of the perversion of design for *use*.

NEED: Much recent design has satisfied only evanescent

wants and desires, while the genuine needs of man have often been neglected by the designer. The economic, psychological, spiritual, technological, and intellectual needs of a human being are usually more difficult and less profitable to satisfy than the carefully engineered and manipulated "wants" inculcated by fad and fashion.

In a highly mobile, throw-away society, the psychological need for security and permanence is often viciously exploited by manufacturer, advertising agency, and salesman by turning the consumer's interest onto the superficial trappings of a transitory "in-group."

People seem to prefer the ornate to the plain as they prefer day-dreaming to thinking and mysticism to rationalism. As they seek crowd pleasures and choose widely traveled roads rather than solitude and lonely paths, they seem to feel a sense of security in crowds and crowdedness. *Horror vacui* is horror of inner as well as outer vacuum.

The need for security-through-identity has been perverted into role-playing. The consumer, unable or unwilling to live a strenuous life, can now act out the role by appearing caparisoned in Naugahyde boots, pseudo-military uniforms, voyageur's shirts, little fur jackets, and all the other outward trappings of Davy Crockett, Foreign Legionnaires, and Cossack Hetmans. (The apotheosis of the ridiculous: a "be-your-own-Paul-Bunyan-kit, beard included," neglecting the fact that Paul Bunyan is the imaginary creature of an advertising firm early in this century.) The furry parkas and elk-hide boots are obviously only role-playing devices, since climatic control makes their real use redundant anywhere except, possibly, in Bismarck, North Dakota.

A short ten months after the Scott Paper Company introduced disposable paper dresses for 99¢, it was possible to buy throw-away paper dresses ranging from $20 to $149.50. With increased consumption, the price of the 99¢ dress could have dropped to 40¢. And a 40¢ paper dress is a good idea. Typically, industry perverted the idea and chose to ignore an important need-fulfilling function of the design: disposable dresses inexpensive enough to make disposability economically feasible for the consumer.

Greatly accelerated technological change has been used

to create technological obsolescence. This year's product often incorporates enough technical changes to make it really superior to last year's offering. The economy of the market place, however, is still geared to a static philosophy of "purchasing-owning" rather than a dynamic one of "leasing-using," and price policy has not resulted in lowered consumer cost. If a television set, for instance, is to be an every-year affair, rather than a once-in-a-lifetime purchase, the price must reflect it. Instead, the real values of real things have been driven out by false values of false things, a sort of Gresham's Law of Design.

As an attitude, "Let them eat cake" has been thought of as a manufacturer's basic right. And by now people, no longer "turned on" by a loaf of bread, can differentiate only between frostings. Our profit-oriented and consumer-oriented Western society has become so overspecialized that few people experience the pleasures and benefits of full life, and many never participate in even the most modest forms of creative activity which might help to keep their sensory and intellectual faculties alive. Members of a "civilized" community or nation depend on the hands, brains, and imaginations of experts. But however well trained these experts may be, unless they have a sense of ethical, intellectual, and artistic responsibility, then morality and an intelligent, "beautiful," and elegant quality of life will suffer in astronomical proportions under our present-day system of mass production and private capital.

TELESIS: "The deliberate, purposeful utilization of the processes of nature and society to obtain particular goals" (American College Dictionary, 1961). The telesic content of a design must reflect the times and conditions that have given rise to it, and must fit in with the general human socio-economic order in which it is to operate.

The uncertainties and the new and complex pressures in our society make many people feel that the most logical way to regain lost values is to go out and buy Early American furniture, put a hooked rug on the floor, buy ready-made phony ancestor portraits, hang a flint-lock rifle over the fireplace, and vote for Ronald Reagan. The gas-light so popular in our subdivisions is a dangerous and senseless anachronism that only reflects an insecure striving for the "good old days" by consumer and designer alike.

Our twenty-year love affair with things Japanese—Zen Buddhism, the architecture of the Ise Shrines and Katsura Imperial Palace, haiku poetry, Hiroshige and Hokusai blockprints, the music of koto and samisen, lanterns and sake sets, green tea liqueur and sukiyaki and tempura—has triggered an intemperate demand by consumers who disregard telesic aptness.

By now it is obvious that our interest in things Japanese is not just a passing fad or fashion but rather the result of a major cultural confrontation. As Japan was shut off for nearly two hundred years from the Western world under the Tokugawa Shogunate, its cultural expressions flourished in a pure (although somewhat inbred) form in the imperial cities of Kyoto and Edo (now Tokyo). The Western world's response to an in-depth knowledge of things Japanese is comparable only to the European reaction to things classical, which we are now pleased to call the Renaissance. Nonetheless, it is not possible to translate things from one culture to another.

The floors of traditional Japanese homes are covered by floor mats. These mats are 3 × 6 feet in size and consist of rice straw closely packed inside a cover of woven rush. The long sides are bound with black linen tape. While tatami mats impose a module (homes are spoken of as six-, eight-, or twelve-mat homes), their primary purposes are to absorb sounds and to act as a sort of wall-to-wall vacuum cleaner which filters particles of dirt through the woven surface and retains them in the inner core of rice straw. Periodically these mats (and the dirt within them) are discarded, and new ones are installed. Japanese feet encased in clean, sock-like tabi (the sandal-like street shoe, or *geta*, having been left at the door) are also designed to fit in with this system. Western-style leather-soled shoes and spike heels would destroy the surface of the mats and also carry much more dirt into the house. The increasing use of regular shoes and industrial precipitation make the use of tatami

difficult enough in Japan and absolutely ridiculous in the
United States, where high cost makes periodic disposal and
re-installation ruinously expensive.

But a tatami-covered floor is only part of the larger design
system of the Japanese house. Fragile, sliding paper walls
and tatami give the house definite and significant acoustical
properties that have influenced the design and development
of musical instruments and even the melodic structure of
Japanese speech, poetry, and drama. A piano, designed for
the reverberating insulated walls and floors of Western
homes and concert halls, cannot be introduced into a
Japanese home without reducing the brilliance of a Rach-
maninoff concerto to a shrill cacophony. Similarly, the
fragile quality of a Japanese samisen cannot be fully appreciated in the reverberating box that constitutes the American house. Americans who try to couple a Japanese interior
with an American living experience in their search for
exotica find that elements cannot be ripped out of their
telesic context with impunity.

Some products, unlike the tatami, are developed without
regard for their aptness. The American automobile is an
example. It is built for roads not yet constructed, super-
fuels still unrefined, and drivers with fast reaction speeds
(yet unborn).

ASSOCIATION: Our psychological conditioning, often going
back to earliest childhood memories, comes into play and
predisposes us, or provides us with antipathy against a
given value.

Increased consumer resistance in many product areas
testifies to design neglect of the associational aspect of the
function complex. After two decades, the television set industry, for instance, has not yet resolved the question of
whether a television set should carry the associational values
of a piece of furniture (a lacquered mah-jongg chest of the
Ming Dynasty) or of technical equipment (a portable tube
tester). Television receivers that carry new associations
(sets for children's rooms in bright colors and materials,
enhanced by tactilely pleasant but non-working controls and
pre-set for given times and channels, clip-on swivel sets for

hospital beds, etc., etc.) might not only clear up the astoundingly large back inventory of sets in warehouses, but also *create* new markets. The tiny transistor pocket radio is still stuck on the threshold of the *trichotomy:* "camera?" "jewelry?" "purse?"; which may account for the monotonous sameness of the many models now on view. While it is eminently possible to tailor these pocket radios to the differing needs and wants of various socio-economic groups and thus force new acceptance areas into being, as it is, transistor radios are still selling almost exclusively to the teen-age crowd.

And what shape is most appropriate to a vitamin bottle: a candy jar of the Gay Nineties, a perfume bottle, or a "Danish modern" style salt shaker?

The response of many designers has been like that so unsuccessfully practiced by Hollywood: the public has been pictured as totally unsophisticated, possessed of neither taste nor discrimination. A picture emerges of a moral weakling with an IQ of about 70, ready to accept whatever specious values the unholy trinity of Motivation Research, Market Analysis, and Sales have decided is good for him. In short, the associational values of design have degenerated to the lowest common denominator, determined more by inspired guesswork and piebald graphic charts rather than by the genuinely felt wants of the consumer.

Since mass production and automation impose both volume and great similarity of types, it becomes increasingly important to extend two basic design approaches:

1. A clear-cut decision as to what the meaning of an object should be. Is an automobile, for instance, a piece of sports equipment, transportation, a living-room-cum-bordello on wheels, or a chrome-plated marshmallow predesigned to turn itself into a do-it-yourself coffin?

2. Greater variety of product sub-types. Can we really expect a Fuller Brush salesman, a university president, a shipyard welder, or a Westchester garden club matron to buy television sets that are identical except for surface finish and the fabric of the speaker grille?

Fixing the product image will demand greater research into the basics of color, form, tactility, visual organization, etc., as well as greater insight into man's perceptual sets and his self-image. The synthesis of these studies, coupled with a greater degree of empathy on the part of the designer, should provide products truer to their intrinsic meaning.

Many products already successfully embody values of high associational content, either accidentally or "by design."

The Sucaryl bottle by Raymond Loewy Associates for Abbott Laboratories communicates both table elegance and sweetening agent without any suggestion of being medicine-like.* The Lettera 22 portable typewriter by Olivetti establishes an immediate aura of refined elegance, precision, extreme portability, and business-like efficiency, while its two-toned carrying case of canvas and leather connotes "all-climate-proof."

Abstract values can be communicated directly to everyone, and this can be simply demonstrated.

If the reader is asked to choose which one of the figures below he would rather call *Takete* or *Maluma* (both are words devoid of all meaning in any known language), he

Figure 2 :   GESTALT COMPARISON

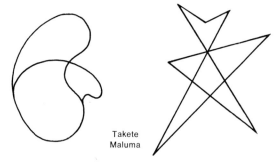

Takete
Maluma

* To this, one must add that *other* bottles manufactured by Abbott Laboratories proved to be insufficiently medicine-like: as of March, 1971, Abbott's infected intravenous-fluid bottles for hospitals had killed nine persons.

will easily call the one on the right *Takete* (W. Koehler, *Gestalt Psychology*).

Many associational values are really universal, providing for unconscious, deep-seated drives and compulsions. Even totally meaningless sounds and shapes can, as demonstrated, mean the same thing to all of us. The unconscious relationship between spectator expectation and the configuration of the object can be experimented with and manipulated. This will not only enhance the "chair-ness" of a chair, for instance, but also load it with associational values of, say, elegance, formality, portability, or what-have-you.

AESTHETICS: Here dwells the traditionally bearded artist, mythological figure with a myth, equipped with sandals, mistress, garret, and easel, pursuing his dream-shrouded designs. The cloud of mystery surrounding aesthetics can (and should be) dispelled. The dictionary definition, *"a theory of the beautiful, in taste and art,"* leaves us not much better off than before. Nonetheless we know that aesthetics is a tool, one of the most important ones in the repertory of the designer, a tool that helps in shaping his forms and colors into entities that move us, please us, and are beautiful, exciting, filled with delight, meaningful.

Because there is no ready yardstick for the analysis of aesthetics, it is simply considered to be a personal expression fraught with mystery and surrounded with nonsense. We "know what we like" or dislike and let it go at that. Artists themselves begin to look at their productions as auto-therapeutic devices of self-expression, confuse license and liberty, and forsake all discipline. They are often unable to agree on the various elements and attributes of design aesthetics. If we contrast the "Last Supper" by Leonardo da Vinci with an ordinary piece of wallboard, we will understand how both operate in the area of aesthetics. In the work of so-called "pure" art, the main job is to operate on a level of inspiration, delight, beauty, catharsis . . . in short, to serve as a propagandistic communications device for the Holy Church at a time when a largely pre-literate population was exposed to a few non-verbal stimuli. But the "Last

Supper" also had to fill the other requirements of function; aside from the spiritual, its *use* was to cover a wall. In terms of *method* it had to reflect the material (pigment and vehicle), tools (brushes and painting knives), and processes (individualistic brushwork) employed by Leonardo. It had to fulfill the human *need* for spiritual satisfaction. And it had to work on the *associational* and *telesic* plane, providing reference points from the Bible. Finally, it had to make identification through association easier for the beholder through such clichés as the racial type, garb, and posture of the Saviour.

Earlier "Last Supper" versions, painted during the sixth and seventh centuries, saw Christ *lying* or reclining in the place of honor. For nearly a thousand years, the well-mannered did not *sit* at the table. Leonardo da Vinci disregarded the reclining position followed by earlier civilizations and painters for Jesus and the Disciples. To make the "Last Supper" acceptable to Italians of his time, on an associational plane, Leonardo sat the crowd around the last supper table on chairs or benches in the proper positions of his (Leonardo's) time. Unfortunately the scriptural account of St. John resting his head on the Saviour's bosom presented an unsolvable positioning problem to the artist, once everybody was seated according to the Renaissance custom.

On the other hand, the primary use of wallboard is to

"The Last Supper," by Leonardo da Vinci.

cover a wall. But an increased choice of textures and colors applied by the factory shows that it, too, must fulfill the *aesthetic* aspect of function. No one argues that in a great work of art such as the "Last Supper," prime functional emphasis is aesthetic, with *use* (to cover a wall) subsidiary. The main job of wallboard is its use in covering a wall, and the aesthetic assumes a highly subsidiary position. But both examples must operate in all *six* areas of the function complex.

We can "feed" any designed object through the filter of our six-sided function complex and understand it better. We can take a certain cologne bottle and examine it on the same basis as the "Last Supper." We will find that it works as well: *aesthetically* we have clear, general form. The surface color used is aquamarine blue ("Aquamarine" is also the name of the cologne), which contrasts well with the shiny brass top. Texture is soft and feels pleasantly yielding to the hand, proportion is excellent, giving the whole bottle unity-with-variety. Satisfying aspects of rhythm, repetition, proportion, and balance can also be observed in the size, placement, and type-face of the lettering. If we continue to the other five aspects of function, we find that the bottle reflects *method* through the good use of materials, tools, and processes. It reflects both the material (glass dipped in vinyl) and the process employed in machine-blown glass. It doesn't topple, and the vinyl coating protects the bottle to some extent from breakage, so that the *use* factor is well satisfied. It answers *need* areas: biological (sex), economic, and psychological (status). Hidden biological needs have been answered by the overall phallic shape. This also operates in the field of *association*. Further examples of the associational field: the color and the name of the cologne are identical; the color moreover has the fragility of the fragrance expected of it. The burnished brass top looks expensively golden; sophistication is powerfully denoted by the use of a script type-face. Its *telesic* content certainly fits some current myths that women are flower-like, expensive sexual playthings.

Designers often attempt to go beyond the primary functional requirements of *method, use, need, telesis, association,* and *aesthetics;* they strive for a more concise statement: precision, simplicity. In a statement so conceived, we find a degree of aesthetic satisfaction comparable to that found in the logarithmic spiral of a chambered nautilus, the ease of a seagull's flight, the strength of a gnarled tree trunk, the color of a sunset. The particular satisfaction derived from the simplicity of a thing can be called *elegance.* When we speak of an "elegant" solution, we refer to something consciously evolved by men which reduces the complex to the simple:

> Euclid's Proof that the number of primes is infinite, from the field of mathematics, will serve: "Primes" are numbers which are not divisible, like 3, 17, 23, etc. One would imagine as we get higher in the numerical series, primes would get rarer, crowded out by the ever-increasing products of small numbers, and that we would finally arrive at a very high number which would be the highest prime, the last numerical virgin.

> Euclid's Proof demonstrates in a simple and elegant way that this is not true and that to whatever astronomical regions we ascend, we shall always find numbers which are not the product of smaller ones but are generated by immaculate conceptions, as it were. Here is the proof: assume that P is the hypothetically highest prime; then imagine a number equal to $1 \times 2 \times 3 \times 4 \ldots \times P$. This number is expressed by the numerical symbol (P!). Now add to it 1: (P! + 1). This number is obviously not divisible by P or any number less than P (because they are all contained in (P!); hence (P! + 1) is either a prime higher than P or it contains a prime factor higher than P . . . Q.E.D.

The deep satisfaction evoked by this proof is aesthetic as well as intellectual: a type of enchantment with the near-perfect.

2

PHYLOGENOCIDE:

A History of the

Industrial Design Profession

We are all in the gutter, but some of us are looking at the stars.

OSCAR WILDE

THE ULTIMATE JOB of design is to transform man's environment and tools and, by extension, man himself. Man has always tried to change himself and his surroundings, but only recently have science, technology, and mass production made this more nearly possible. We are beginning to be able to define and isolate problems, to determine possible goals and work meaningfully towards them. And an over-technologized, sterile, and inhuman environment has become one possible future; a world choking under a permanent, dun-colored pollution umbrella, another. In addition the various sciences and technologies have become woefully compartmentalized and specialized. Often, more complex problems can be attacked only by teams of specialists, who often speak only their own professional jargon. Industrial designers, who are often members of such a team, frequently find that, besides fulfilling their normal design function, they must act as a communication bridge between other team members. Frequently the designer may be the only one who speaks the various technical jargons. Because of his educational background, the role of team interpreter is often forced upon him. So we find the industrial designer in a team situation becoming the "team synthesist," a position to which he has been elevated only by the default of people from all the other disciplines.

This has not always been true.

Many books on industrial design suggest that design began when man began making tools. While the difference between *Australopithecus africanus* and the modern designer may not be as great as one might think or hope for, the stance of equating man the toolmaker with the start of the profession is just an attempt to gain status for the profession by evoking a specious historical precedent. "In The Beginning Was Design": obviously, but not industrial design. Henry Dreyfuss, one of the founders of the profession, says in *Designing for People* (probably the best and most characteristic book about industrial design):

> The Industrial Designer began by eliminating excess decoration, but his real job began when he insisted on dissecting the product, seeing what made it tick, and devising means of making it tick better—then making it look better. He never forgets that beauty is only skin-deep. For years in our office we have kept before us the concept that *what we are working on is going to be ridden in, sat upon, looked at, talked into, activated, operated, or in some way used by people individually or en masse. If the point of contact between the product and the people becomes a point of friction, then the Industrial Designer has failed. If, on the other hand, people are made safer, more comfortable, more eager to purchase, more efficient—or just plain happier—the designer has succeeded.* He brings to this task a detached, analytical point of view. He consults closely with the manufacturer, the manufacturer's engineers, production men, and sales staff, keeping in mind whatever peculiar problems the firm may have in the business or industrial world. He will compromise up to a point but he refuses to budge on design principles he knows to be sound. Occasionally he may lose a client, but he rarely loses the client's respect.

Industrial design, then, is always related to production

and/or manufacturing facilities, a state of affairs enjoyed by neither man nor the Deity.

The first concern with the design of tools and machinery coincided almost exactly with the beginnings of the industrial revolution and, appropriately enough, made its first appearance in England. The first industrial design society was formed in Sweden in 1849, to be followed shortly by similar associations in Austria, Germany, Denmark, England, Norway, and Finland (in that order). The designers of the period were concerned with form-giving, an erratic search for "appropriate beauty" in machine tools and machine-made objects. Looking at the machine, they saw a new thing, a thing that seemed to cry out for decorative embellishments. These decorations were usually garnered from classical ornaments and from major raids into the animal and vegetable kingdoms. Thus, giant hydraulic presses dripped with acanthus leaves, pineapples, stylized wheat sheaves. Many of the "sane design" or "design reform" movements of the time, as those engendered by the writings and teachings of William Morris in England and Elbert Hubbard in America, were rooted in a sort of Luddite anti-machine philosophy. Frank Lloyd Wright said in 1894 that "the machine is here to stay" and that the designer should "use this normal tool of civilization to best advantage instead of prostituting it as he has hitherto done in reproducing with murderous ubiquity forms born of other times and other conditions which it can only serve to destroy." Yet designers of the last century were either perpetrators of voluptuous Victorian-Baroque, or members of an artsy-craftsy clique who were dismayed by machine technology. The work of the *Kunstgewerbeschule*, in Austria, and some isolated German design groups anticipated things to come, but it was not until Walter Gropius founded the German Bauhaus in 1919, that an uneasy marriage between art and machine was achieved.

No design school in history has had greater influence in shaping taste and design than the Bauhaus. It was the first school to consider design a vital part of the production process rather than "applied art" or "industrial arts." It became the first international forum on design because it drew its faculty and students from all over the world, and its influence traveled as these people later founded design offices and schools in major countries. With true Teutonic thoroughness, the Bauhaus attempted to evolve a methodology of design and design teaching. Almost every major design school in the United States today still uses the basic foundation design course developed by the Bauhaus. It made sense in 1919 to let a German nineteen-year-old experiment with drill press and circular saw, welding gun and lathe, so that he might "experience the interaction between tool and material." Today the same method is an anachronism, for an American teen-ager has spent his entire life in a machine-dominated society (and cumulatively probably a great deal of time lying under various automobiles, souping them up). For a student whose American design school still slavishly imitates teaching patterns developed by the Bauhaus, computer sciences and electronics and plastics technology and cybernetics and bionics simply do not exist. The courses which the Bauhaus developed were excellent for their time and place (telesis), but American schools following this pattern in the seventies are perpetuating design infantilism.

The Bauhaus was in a sense a non-adaptive mutation in design, for the genes contributing to its convergence characteristics were badly chosen. In bold-face type, it announced its manifesto: "Architects, sculptors, painters, we must all turn to the crafts. . . . Let us create a *new guild of craftsmen!*" The heavy emphasis on interaction between crafts, art, and design turned out to be a blind alley. The inherent nihilism of the pictorial arts of the post-World War I period had little to contribute that would be useful to the average, or even to the discriminating, consumer. The paintings of Kandinsky, Klee, Feininger, et al., on the other hand, had no connection whatsoever with the anemic elegance that some designers imposed on products.

In America, industrial design, like marathon dances, six-day bicycle races, the NRA and the Blue Eagle, and free dishes at the movies, was a child of the Depression. At first glance the swollen belly of a child suffering from malnutrition gives it the appearance of being well fed; later you notice the emaciated arms and legs. The products of early American industrial design convey the same sleek obesity and have the same weaknesses.

For the Depression market, the manufacturer needed a new sales gimmick, and the industrial designer reshaped his products for better appearance and lower manufacturing and sales costs. Harold Van Doren's definition of that time in *Industrial Design* was apt:

> Industrial Design is the practice of analyzing, creating, and developing products for mass-manufacture. Its goal is to achieve forms which are assured of acceptance before extensive capital investment has been made, and which can be manufactured at a price permitting wide distribution and reasonable profits.

Harold Van Doren, Norman Bel Geddes, Raymond Loewy, Russel Wright, Henry Dreyfuss, Donald Deskey, and Walter Dorwin Teague were the pioneering practitioners of design in America. It is significant that all of them came from the field of stage design and/or window display.

While the architects sold apples on street corners, the ex-stage-designers and ex-window-dressers were creating "lemons" in lush suites upstairs.

Raymond Loewy's redesign of the Gestetner duplicating machine is probably the first and most famous case of industrial design development. But as Don Wallance was to remark three decades later in *Shaping America's Products:*

> The "before and after" pictures showing Mimeographs, locomotives, refrigerators, furniture and numerous other things transformed by industrial design were most impressive. Even more impressive were the differences in before and after sales figures. Oddly enough, when we look at these things now after a passage of more than twenty-five years, it is no longer so clear whether the "before" or the "after" version has best stood the test of time.

This sort of design for the manipulated visual excitement

of the moment continued unabated until the beginning of World War II.

The automobile and other consumer industries had to turn their production facilities over to the creation of war supplies, and war-time demands forced a new (though temporary) sense of responsibility on industrial designers. "Ease-o-matic gear shifts" and "auto-magic shell feeding mechanisms" were out of place in a Sherman tank. Design staffs encountered real requirements of performance in the function complex, imposed by combat conditions. The necessity for honest design (design-in-use versus design-in-sales) imposed a healthier discipline than that of the market place. Critical material shortages forced these designers who remained in the consumer field to a much keener realization of performance, materials, and other war-imposed limitations. A three-quart casserole, made of plasticized cardboard able to sustain temperatures of 475° for several hours, washable and infinitely reuseable, *retailing* for 45¢, is an excellent example and seems curiously to have disappeared from the market at the end of 1945.

During the war the societal influences on design in the United States fell into three different distinct stages. The first was: "We must kill Nazis because they kill Jews and want to conquer the world, and that is bad," or the "it's a job; let's get it over with" phase. The message was: austerity and commitment. Within the function complex the designer had to stress *use* and *need*. The second phase was: "We're fighting this war for Mom, blueberry pie, the girl next door, and Nash-Kelvinator" or "I hear America singing." The message was: determination and nostalgia. Designs seemed to reflect a preoccupation with *method* and *telesis*. The third stage was: "When it's all over, you can land your own 'copter on top of your plasticized, climate-controlled pleasure dome, complete with wall-to-wall tropical fish tanks, bartender-robots, and a tri-di video set" or the "we've made the world safe for technocracy" phase. The message was: fulfillment and hedonism. Design succumbed to an orgiastic abandonment in *associational* and *aesthetic* areas.

Shortly after the end of the war *The New York Times* carried the first full-page ad for Gimbel's sale of the Reynolds ball-point pen at only $25 each. By Monday morning, Herald Square was so clogged with people waiting for Gimbel's to open that extra police had to be brought in to control the crowds. Places in the ball-point pen queue could be sold for $5 to $10, and until Gimbel's slapped a one-pen-to-a-customer order on the sale Wednesday, pens could be readily sold for $50 to $60 each.

This zany state of affairs lasted for some five weeks: each day Hudson Lodestar monoplanes landed at LaGuardia with thousands of pens in their bellies. Even a three-day truckers' strike couldn't affect the sale, as the union promised to "deliver milk, critical food and Reynolds pens." With a Reynolds pen you could "write under water" but practically nowhere else. They skipped, they blotted, they leaked in your pockets, and there were no replacement cartridges because the pens were one-shot affairs. You threw them away as soon as they ran dry, if not sooner. Still they sold. For the pen was a do-it-yourself Buck Rogers kit; you bought a pen and you were "post-war"; just as every "ruptured duck" dully gleaming in the lapel of a service man's first civilian suit marked the end of one era, the Reynolds pen leaking in his breast pocket marked the beginning of a new. There were other consumer goods available, but this was the only totally new product on the market.

The technology of the year 2000 had come to roost in 1945. Each man's miraculously lightweight Reynolds pen gleaming in aluminum was also his personal reassurance that "our side" had won the war.

(Now it can be told: "our" pen was copied from German ball-points found by Reynolds in a South American bar in 1943.)

Industry pandered to the public's ready acceptance of anything new, anything different. The miscegenative union between technology and artificially accelerated consumer whims gave birth to the dark twins of styling and obsolescence. There are three types of obsolescence: technological (a better or more elegant way of doing things is discovered), material (the product wears out), and artificial (the death-rating of a product; either the materials are substandard and will wear out in a predictable time span, or else significant parts are not replaceable or repairable). Since World War II our major commitment has been to stylistic and artificial obsolescence. (Ironically enough, the accelerated pace of technological innovation frequently makes a product obsolete before artificial or stylistic obsolescence can be tacked onto it.)

In the seventies the social environment within which design operates has undergone still another change, since society itself has become more polarized. Within the United States the "poor are getting poorer" and the fat cats are getting incredibly fatter. On one hand, the middle class attempts to express itself more and more through possession of "campy" little gadgets and to find both identity and value through owning products. On the other hand, abject poverty (previously, decorously and piously hidden away like a demented maiden aunt in a nineteenth-century New England attic) has emerged as a major reality of life. There *are* children starving to death in parts of Mississippi and South Carolina. There are vast population sinks in the large city ghettos, the inhabitants of which share absolutely none of the motivations and aspirations of the middle class. And by no means are all of these people of Negro, Puerto Rican, or Mexican descent. Embittered "senior citizens" from our rural areas, foiled in their attempts to "retire gracefully on $150 a month at age 65," haunt the sleazy resorts of Florida, southern Texas and southern California, dreaming a paranoid dream of fascist restoration to the "good old days."

On a global scale the disparities between the haves and have-nots have become even more terrifying vast. Averaging things out from R. Buckminster Fuller's *Inventory of World Resources*, we find that per capita ownership of "energy slaves" shows this trend most clearly.

An "energy slave" is determined as follows: in addition to the energy spent from the metabolic income in "working" his own body, one man in one 8-hour day can do

approximately 150,000 foot-lbs. of work. A foot-lb. of
our work equals about the amount of energy required
to lift one pound one foot vertically. This additional
work might be called net advantage in dealing with an
environment. The "net advantage" potentially to be
gained by each human, each year, working 8 hours
each of 250 days per year, is 37.5 million foot-lbs.

Stated with a probable error of less than 10%, the
world consumption of energy from mineral fuels (coal,
oil, gas) and water power, from the year 1960, is 94.4
quintillion (94,429,000,000,000,000,000) foot-lbs. Assuming man's efficiency in converting his gross energy
consumption into work to average an overall 4%, he
will net 3.7 quintillion (3,769,960,000,000,000,000)
foot-lbs.

Dividing this figure by 37.5 million foot-lbs. (each man's
net annual energy advantage), we find that a 100.6
billion man-year equivalence of work is being done for
him. The 100.6 billion man-equivalence we will call
100.6 billion energy slaves.

$$\frac{100.6 \text{ million energy slaves}}{2.995 \text{ billion work population}} = \frac{33.5 \text{ energy slaves}}{\text{per capita.}}$$

However, these energy slaves were not divided up
equally on the face of the earth. Marked contrasts are
to be seen, e.g., each of the 199 million "North American" inhabitants is served by 185 slaves (460 slaves per
family) while each of the inhabitants of Asia is now
limited to the services of 3 slaves. (1960)

Since 1960 this chasm has widened. With the declining
birth rate in both the North American and the Western
European technate and the fantastic population explosion
in the rest of the world, the balance sheet now shows 208
energy slaves per North American and 0.9 energy slaves
per inhabitant in the so-called "emergent nations."

By extrapolating Fuller's figures and adding UNESCO
statistics we find that finally the correlation between the
"primary useful product life" and the actual length of use
also works to the disadvantage of the have-not parts of the
world, as the following brief table will show:

PRODUCT	PRIMARY USEFUL PRODUCT LIFE IN YEARS	ACTUAL TIME USED IN U.S. IN YEARS	ACTUAL TIME USED IN UNDER-DEVELOPED COUNTRIES IN YEARS
Bicycle	25	2	75
Washing Machines & Irons	5	5	25
Band-Power Tools	10	3	25
Automobiles	11	2.2	40+
Construction Equipment	14	8	100+
General Purpose Industrial Equipment	20	12	75+
Agricultural Machinery	17	15	2500+
Railroad Equipment	30	30	50
Ships	30	15	80+
Miniaturized Hi-Fi, Photographic & Film Equipment	35	1.1	50

So it seems that the energy slaves (encapsulated in equipment) we *do* have, we manage to fritter away.

Much of the above may lie squarely in the areas of sociology and economics, but, as previously pointed out, industrial design attempts a horizontal synthesis that cuts across narrow disciplines and narrow minds. Even though in nearly all other disciplines there is a greater emphasis on vertical specialization, the above is of legitimate concern to the designer (or should be).

In *Never Leave Well Enough Alone*, Raymond Loewy amusingly reminisces about the early years of his crusade, a crusade to get clients. In the late twenties and thirties he, together with other designers, kept knocking on corporate doors, such as those of General Motors, General Electric, General Rubber, General Steel, General Dynamics. In all fairness it must be admitted that he and his co-workers served their corporate masters well and, in fact, still do. But it is dismaying to find that all too many of today's graduating design students eagerly join corporate design staffs, safely wrapped in the cocoon of corporate expense accounts, company-paid country club memberships, deferred annuities, retirement benefits, dread-disease coverage, and a yearly revitalizing visit to one of the stamping grounds of the corporation gauleiters in New England or Aspen, Colorado.

By now it is abundantly clear that a new crusade is needed. Vast areas of need, and concomitantly, need for design, exist all over the world. It is up to the designer (like Raymond Loewy, but in a socially and morally more acceptable way) to knock on doors that have never opened before.

The choice is *not* between corporate charcoal gray, buttoned-down security, on one hand, and a giggling freak-out, high on LSD in a Haight-Ashbury gutter, on the other. There is a third way. The Office of Economic Opportunity, the Southern Appalachian Project, the International Labor Organization in Geneva, Switzerland, UNESCO, UNICEF, as well as many other organizations (of various political colorations) in hundreds of areas concerned with optimal human survival needs: these are just a few directions in which designers should and must move.

3

THE MYTH OF

THE NOBLE SLOB:

Design, "Art," and the Crafts

Good Taste is the most obvious resource of the insecure.
People of good taste eagerly buy the Emperor's old clothes.
Good taste is the first refuge of the noncreative.
It is the last-ditch stand of the artist.
Good taste is the anesthetic of the public.

HARLEY PARKER

THE CANCEROUS GROWTH of the creative individual expressing himself egocentrically at the expense of spectator and/or consumer has spread from the arts, overrun most of the crafts, and finally reached even into design. No longer does the artist, craftsman, or in some cases the designer, operate with the good of the consumer in mind; rather, many creative statements have become highly individualistic, auto-therapeutic little comments by the artist to himself. As early as the mid-twenties there appeared on the market chairs, tables, and stools designed in Holland by Wijdveldt, as a result of the *De Stijl* movement in painting. These square abstractions painted in shrill primaries were almost impossible to sit in; they were extremely uncomfortable. Sharp corners ripped clothing, and the entire zany construction bore no relation to the human body. Today we can allow ourselves to sneer at these attempts to transfer the two-dimensional paintings of Piet Mondrian and Theo van Doesburg into "home furnishing." The chairs lasted as sophisticated status symbols for only a few years, but the trend, the attempt to translate fashionable daubs into three-dimensional objects for daily use, still continues. Salvador Dali's sofa constructed in the shape of Mae West's lips may have been a "disengaged" surrealist act, much like Meret Oppenheim's fur-lined cup and saucer, but the oppy-poppy pillows of today sell by the thousands. While it is not a bad idea to have a pillow that sells for $1.50, a pillow that can be folded and stored in one's watch pocket and blown up for use, these small plastic horrors perform none of their functions. They yield but slightly and, being made of transparent plastic with silk-screened polka dots, don't "breathe"— hence the user sweats profusely. In pictures in shelter magazines (such as *House Beautiful* and *House and Garden*), these pillows are usually shown in clusters, but when several are placed together they have an unfortunate tendency to *sqeak* against one another like suckling pigs put to the knife. A clutch of these pillows, tastefully "unarranged" on a day bed, looks pleasant. Their use indicates that, as in so many other areas, we have sacrificed all our other needs for a purely visual statement. As the pillows are also bought in a purely visual setting, dissatisfaction does not set in until one attempts to use them. And imagine the dismay of having some romantic interlude punctured by a sudden pillow blow-out.

With new processes and an endless list of new materials at his disposal, the artist, craftsman, and designer now suffers from the tyranny of absolute choice. When everything becomes possible, when all the limitations are gone, design and art can easily become a never-ending search for novelty, and the desire for novelty on the part of the artist becomes an equally strong desire for novelty on the part of the spectator and consumer, until newness-for-the-sake-of-newness becomes the only measure. It is at this point that many different versions of novelty begin to create many different esoteric consumer cliques, and the designer with his wares may become more and more alienated from his society and from the functional complex.

In his novel *Magister Ludi,* Hermann Hesse writes about

a community of intellectual elites who have perfected a mystical, symbolic language, called the "Bead Game," that has reduced all knowledge to a sort of unified field theory. The world outside the community is convulsed by riots, wars, and revolutions, but the players of the Bead Game have lost all contact. They are engaged in exchanging their esoterica with one another in the game. There is a disturbing parallel between Hesse's game and the aspirations of the contemporary artist when he speaks of his goals in the exercise of his private visions. He discourses on space, the transcendence of space, the multiplication of space, the division and negation of space. It is a space devoid of man, as though mankind did not exist. It is, in fact, a version of the Bead Game.

Concerning the artist Ad Reinhardt, *Time* says:

> Among the new acquisitions currently on display at Manhattan's Museum of Modern Art is a large square canvas called "Abstract Painting" that seems at first glance to be entirely black. Closer inspection shows that it is subtly divided into seven lesser areas. In a helpful gallery note at one side, Abstractionist Ad Reinhardt explains his painting. It is: "A square (neutral, shapeless) canvas, five feet wide, five feet high, as high as a man, as wide as a man's outstretched arms (not large, not small, sizeless), trisected (no composition), one horizontal form negating one vertical [sic] form (formless, no top, no bottom, directionless), three (more or less) dark (lightless), non-contrasting, (colorless) colors, brushwork brushed out to remove brushwork, a mat, flat freehand painted surface (glossless, textureless, non-linear, no hard edge, no soft edge) which does not reflect its surroundings—a pure, abstract, non-objective, timeless, spaceless, changeless, relationless, disinterested painting—an object that is self-conscious (not unconsciousness), ideal, transcendent, aware of nothing but art (absolutely no anti-art)."

This from one of America's "most eloquent artists."

The tomes by learned art historians make a great to-do over the influence of the camera and photography on the plastic arts. And it is certainly true that by placing an apparatus in everyone's hand that made "copying nature" possible for everyone with enough wit to push a button, one of the main objectives of painting—to produce a high-fidelity reproduction—seemed partially fulfilled. It is usually overlooked that even a photograph is a first-order abstraction. Thus in the Galician and Polish backwater sections of the old Austro-Hungarian Empire, village pharmacists did a brisk trade in male model photographs at the beginning of World War I. Each of these wily shopkeepers would stock four stacks of small identical photographs of a cabinet view of male models, 5½ × 4 inches in size. One picture showed the face of a clean-shaven man. The second, that of a man with a moustache. The third picture showed a man with a full beard, while, in the fourth, the model's hirsute elegance encompassed both beard and moustache. A young man called up for military service bought the one of the four photographs that most nearly matched his own face and presented it to his wife or sweetheart to remember him by. And it worked! It worked because the picture of even a stranger with the right kind of moustache *was closer to the face of the departed husband than anything his wife had ever seen before except for his face itself.* (Only by glancing at several, or many, photographs could she have gained the sophistication to be able to differentiate among these various first-order abstractions.) But the role of photography and its influence on art is by now fairly well documented and established.

However, hardly anyone has considered the important impact made by the machine tool and machine perfection. The tolerances demanded of the case for a Zippo cigarette lighter and achieved by automatic handling machinery are far more exact than anything Benvenuto Cellini, possibly the greatest metalsmith of the Renaissance, could have achieved. With modern space hardware technology, plus-minus tolerances of $1/10,000$ of an inch are a routine production achievement. This is not to make a value judgment of Cellini versus an automated turret lathe; it is merely to show that "mere perfection" can be routinely had on assembly lines and in factories, thus depriving the plastic arts of a second goal, the "search for perfection." Like it or not, the contemporary artist lives in contemporary society. Man lives today as much in the environment of the machine as the machine lives in the environment of man. It may be belaboring the obvious to say that there are more man-made objects in the landscape than landscape itself. Unable personally to cope with this change in the environment, the modern artist has created a series of escape mechanisms for himself.

Ineluctably, the artist lives in a technological world. Even an academic landscape painter living in, say, Cornwall, is bound to see more automobiles than cows on any given day. Some artists, then, see the machine as a threat, some as a way of life, some as salvation. All of them have to find a way to live with it.

Seemingly, one simple way to get rid of a threat is to poke fun at it. While, from its earliest days at the *Cabaret Voltaire* in 1916, the Dadaist movement attempted to show the general absurdity of twentieth-century man and his world, there was always a heavy dose of satirization of the machine involved. From Marcel Duchamp's "ready-mades" ("Why not sneeze," "Fountain," etc.) to many of Max Ernst's "collages," to the satirical conglomerations of mass-produced items in Kurt Schwitters's "Merzbau," the attempt has been to make fun of the machine through ridicule, satire, or burlesque. A contemporary manifestation is Jean Tinguely's "machines." These vast constructs of cogs, screws, umbrella guts, pinwheels, light bulbs, and deflowered sewing machines shake, jiggle, and quake, sometimes exploding or (disappointingly) just smoldering a little. In 1960 one of these "sculptures," composed of myriad pieces of old machines, was erected in the garden of the Museum of Modern Art in New York and, with the setting sun, began to grind into motion. To the delectation of an overflow audience, parts of the sculpture frenziedly came into action, catching fire and burning until they collapsed into puddles of kerosene and rust, a proceeding that was viewed with some dismay by the New York City fire companies that were called by frightened neighbors.

Over-compensation can be fun, too. Piet Mondrian, finding himself surrounded by machine-made precision in the middle twenties in Holland, decided to turn himself into a machine. His square white canvases divided by narrow black bands with only two or three primary-colored squares or rectangles dynamically balanced could well be the result of machine production. In fact, a computer in Basel, Switerland, is at present creating Mondrian-like pictures. This may raise the question of creativity; the computer versus Mondrian. Leaving aside the fact that the computer had to be programmed, it shows just what aspect of Mondrian's work was really creative. Shortly after his death, I saw a Mondrian retrospective showing in New York which included some of his unfinished canvases. The black pigment lines were represented by lengths of black tape; on the white background one could still discern the traces of the tape having been moved back and forth. Obviously, Mondrian tried various positions. Having known him during his illness, I know that he would have preferred to sit calmly back in his chair and have two servants move the lines and color areas back and forth until he considered them to be in perfect balance. Had he lived to see graphic readout computers, he would have found them a delightful new toy. From the traces of tape on his white unfinished canvases we can see that Mondrian himself followed computer-like behavior patterns and that what creativity he brought to the process of painting was entirely in the area of aesthetic decision-making. Piet Mondrian's work has found a ready but debased acceptance in the façade design of contemporary buildings, Kleenex packages, and typographical layout.

A third way of dealing with the machine is to avoid it altogether. The Surrealist movement, inheritors of the irrational side of Dadaism, attempted to plumb that region, half cesspool and half garden, known as the unconscious, or id. By basing their highly realistic canvases on subconscious symbols, they hoped to turn themselves into latter-day medicine men, witch doctors, shamans of the pigment. The trouble with this concept is that id-motivated emotions differ from person to person. Salvador Dali may experience a world of voluptuous sexuality from his painting of a burning giraffe (and in fact considers it his most potent painterly sexual stimulus), but it does not communicate sexuality to any of his spectators. Dorothea Tanning's jack-booted naked ten-year-old girl wearing a Gay Nineties sailor hat and sensuously embracing the red-hot pipe of a stove also fails to elicit an appropriate response. There has been a lot of loose talk about the "left hand being the dreamer," Jungian archetypes, intuitional and poetic feeling tones, metaphysics, mysticism, etc. But all the totemistic and fetishistic emblems of the Surrealists failed to come across. A reference point was missing. The Comte de Lautréamont described surrealism as being like "the chance encounter of a sewing machine and an umbrella upon a dissecting table," but thousands of those surreal chance encounters have taken place since then—and some are now one with the hot dust of Spain, of Europe, of Vietnam—and the concept is no longer bizarre.

The human preoccupation of liking to play with doll houses has been used cleverly by Joseph Cornell. His little boxes with strange and esoteric objects cunningly arranged therein are manageable small universes, perfect within themselves, into which no hint of masscult or midcult can enter.

Seeking refuge from the threatening surround by catering to a small coterie (as in the Bead Game) was carried to its highest level by Yves Klein. Some of his methods are described in the book *Collage*. When not busy glueing 426,000 sponges to the wall of a resort hotel, Mr. Klein was fond of painting watercolors and then placing them in his back yard during heavy rains in order to "obtain a dynamic interchange between nature and man-made images." He used the same rationale for doing oil paintings in a slow-drying vehicle and then strapping them to the roof of his Citroën and driving briskly around "to make the colors clarify." The height of his career was reached when the *Galerie Iris Clert* held his first non-painting show in 1958. The gallery had been painted festively in white, the only objects in view were simple white frames hanging on the walls with descriptions of prices attached, such as: "Non-painting, 30 cm. × 73 cm., Fr. 80,000." The show was a sell-out. Hundreds of Parisians and American visitors solemnly paid for and carried empty white frames to their cars, and, one supposes, then hung them triumphantly in their living rooms. It would be instructive to find out if Mr. Klein would have accepted non-checks.

Although Andy Warhol, Roy Lichtenstein, and Robert Rauschenberg have surrounded their productions with much rationale, their attempts to reduce the unusual to the commonplace and raise the commonplace to the stature of the unusual, are losing propositions. Marilyn Monroe's face identically stenciled 50 times merely attempts to say that Miss Monroe was one of a herd, and interchangeably so, a charge that can be leveled at most Hollywood sex symbols but certainly not at Miss Monroe. The reduction of human emotions to the level of a comic strip episode is an attempt to shield oneself from involvement through banality. And Marcel Duchamp said in a recent periodical: "If a man takes fifty Campbell Soup cans and puts them on a canvas, it is not the retinal image which concerns us. What interests us is the kind of mind that wants to put fifty Campbell Soup cans on a canvas."

Art as self-gratification can, of course, also act as an outlet for aggression and hostility. Niki de Saint-Phalle fires rounds of ammunition into her white plaster constructions from a gun, releasing little bags of paint that spurt out and dribble all over her pieces. When not engaged with plaster of Paris and her fowling piece, Miss de Saint-Phalle got together with two "collaborators" and constructed a gigantic reclining nude in Stockholm which spectators enter through the vagina to view the interior constructions, a merry-go-round for the kiddies and a cocktail bar within her generously proportioned mammary glands.

Earlier we spoke about the artist suffering from the tyranny of absolute choice. But if he doesn't care to poke fun at the machine, become a machine, turn himself into a bogus witch doctor, construct tiny boxed universes, elevate the commonplace to a symbol of banality, or let out his aggressions on a middle class no longer capable of being shocked, the area of choice is narrowed abruptly. One thing remains: accidents. For a well-programmed computer makes no mistakes. A well-designed machine is free of error. What, then, is more logical than to glorify the mistakes and to venerate the accidents. Jean (Hans) Arp, one of the co-founders of the Dada movement in Zurich during World War I, tried it first: "Forms Arranged According to the Laws of Chance."

Mr. Arp tore up one of his gouache paintings (without looking) then climbed up on top of a step ladder and let the pieces drop. Carefully, he glued them down where they had fallen. A few decades later, another Swiss, named Spoerri, will invite his girl friend for breakfast, then glue all the dishes, soiled paper napkins, bacon rinds, and cereal dregs down on the table, entitle the result "Breakfast with Marie," and hang it, table and all, in a museum. It was probably unavoidable that after Jackson Pollock's dribble-and-blob paintings of the forties and early fifties, the Abstract Expressionist painters would whoop it up for mistake, accident, and the unplanned. One member of this group paints by strapping his brushes to his left forearm since, he says, he "can't breed the ability even out of his left hand." With other painters rolling naked models across their canvases, or riding across their wet paintings on motorcycles, scooters, bicycles, roller skates, or trampling across them on snow-shoes, the "desire for novelty" is getting full play.

Most of what has been said before regarding the artist's relation to a machine culture still holds true with the most recent movements. To this can be added that, with a continuing search for things that will be perceived as "different," avant garde, or "far out" by his clients or spectators, the contemporary artist has tended to become more "trendy" in his work.

Increasingly, many of us (especially the young) have come to reject material possessions, objects, products per se. That this emotion is engendered largely by the fact that we live in a post-industrial society bursting with gadgets, knick-knacks, and manufactured trivia, is abundantly clear. So now we have "Conceptual Art." A recent production by a leading West Coast painter consisted of some 15 pages of yellow paper. On each page he described with meticulous detail the sizes, colors, textures, and compositions that would have constituted nearly 400 paintings, *provided he had painted them*. Added to this were descriptions of the working conditions under which these canvases would have been painted, had they been painted at all. After reading these descriptive passages, he then burned the papers.

George McKinnon, an exhibition-oriented photographer from the West Coast, photographs pictures appearing in old magazines and entitles these "retrospective pieces."

When museum patrons are invited to a formal opening and are advised *not* to come to the museum but to go instead to the Sixty-third Street subway station and peer into the mirror of a gum-vending machine on the second level; while their friends at exactly the same time are advised to take the Staten Island Ferry and spend the entire trip in the toilet reading *Silence*, by John Cage; and still another party is told to rent a room at the Americana Hotel and spend the time shaving; and all these many activities, indulged in by all these many people simultaneously, constitute both the opening of the art show, as well as the art show itself; we are in the presence of folks trying to play random games. And randomness, as has been stated earlier, is the one game the machine won't play, and, therefore, this too is a reaction against the machine.

Ever since the environment became an "in" thing, we have had Earthworks as another artistic trend. Now Earthworks can be many things: A 30-foot trench dug in the Mojave Desert, one leaf torn off every third oak tree in Tallahassee, Florida; or for that matter, snow lying on a meadow in Colorado, to which nothing has been done whatever.

I don't wish to make any judgments about others finding meaningful creative and artistic engagement by pissing into a snow bank, but surely the good people working in the arts can find more authentic ways of surprising us, delighting us, or reflecting their views.

(Incidentally, all of this and what the future will bring in the arts was listed, described, and explained in a book written in 1948 in England by C. E. M. Joad. Its incredibly appropriate title is *Decadence*.)

What is the relevance of all "art games" of this sort to life? Without question, our time needs paintings, music, sculpture, and poetry. Seldom in fact have both delight and catharsis been needed more. An alternate view might hold that the artists of China, Cuba, and North Vietnam have helped their people's struggle toward liberation through developing new modes of expression that are firmly based on underlying folk art motifs and peasant culture.

Of course, even the self-indulgent salon drivel of New York and San Francisco and Los Angeles can be justified in terms of the people who are doing it. But a recent encounter in New York can at least point to an alternate view: when a number of "painters" smashed two dozen violins and bass fiddles to glue the fragments onto a wall and create a mural, some probing questions were asked about the young Puerto Ricans and Blacks in the neighborhood who might wish to study music but who could never afford to buy instruments. . . .

And where are there permanent design collections? Besides the Museum of Modern Art in New York, there are vestigial collections in Minneapolis, San Francisco, Boston, and Buffalo, N.Y. The rest of the country may sometimes see a traveling "good design" exhibition, but their exposure to well-designed objects ends with that.

And for that matter, even the most prestigious exhibitions of "good design" can be disappointments. In New York the Museum of Modern Art recently held an exhibition of "well-designed" objects that elevated the ugly, in fact *the consciously ugly*, to a new level. Thus we could see a small, high-intensity lamp that has been designed to look precarious and unstable no matter at what angle it is put down. An unruly gush of plastic, colored precisely the shade of frozen diarrhea, doubles as an easy chair. In short, in a society in which the "machine perfect" or even the "fashionably pleasing" can be obtained with a minimum of effort, grossness and the ugly have become imbued with value to the untrained and under-equipped spectator or consumer.

Other such exhibitions of objects are discussed in Chapter Six.

If design is a problem-solving activity, this kowtowing to the lowest common denominator has no reason for existence. It is only when the designer abrogates his responsibilities to himself and others and operates as a pimp for the sales department that he finds this creation of warmed-over "soul food" palatable at all.

Much has been said about the decadence of Rome when the barbarians were outside the gate. There are no barbarians outside ours: we have become our own barbarians, and barbarism has become a do-it-yourself kit.

"Armchair" (1964) by Gunnar Aagaard Andersen. Urethane foam, 30" high. Executed at Dansk Polyether Industri, Denmark. Collection The Museum of Modern Art, New York. Gift of the designer. While the chair is ugly, it is incredibly comfortable and "grown" out of foam biomorphically.

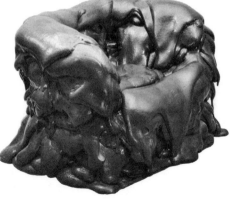

4

DO-IT-YOURSELF
MURDER:
The Social and Moral
Responsibilities of the Designer

The truth is that engineers are not asked to design for safety. Further inaction will be criminal—for it will be with full knowledge that our action can make a difference, that auto deaths can be cut down, that the slaughter on our highways is needless waste . . . it is time to act.

ROBERT F. KENNEDY

ONE OF MY FIRST JOBS after leaving school was to design a table radio. This was shroud design: the design of the external covering of the mechanical and electrical guts. It was my first, and I hope my last, encounter with appearance design, styling, or design "cosmetics." The radio was to be one of the first small and inexpensive table radios to compete on the post-war market. Still attending school part-time, I naturally felt insecure and frightened by the enormity of the job, especially since my radio was to be the only object manufactured by a new corporation. One evening Mr. G., my client, took me out on the balcony of his apartment overlooking Central Park.

He asked me if I realized the kind of responsibility I had in designing a radio for him.

With the glib ease of the chronically insecure, I launched into a spirited discussion of "beauty" at the market level and "consumer satisfaction." I was interrupted. "Yes, of course, there is all that," he conceded, "but your responsibility goes far deeper than that." With this he began a lengthy and cliché-ridden discussion of his own (and by extension his designer's) responsibility to his stockholders and especially his workers.

> Just think what making your radio entails in terms of our workers. In order to get it produced, we're building a plant in Long Island City. We're hiring about 600 new men. Now what does that mean? It means that workers from many states, Georgia, Kentucky, Alabama, Indiana, are going to be uprooted. They'll sell their homes and buy new ones here. They'll form a whole new community of their own. Their kids will be jerked out of school and go to different schools. In their new subdivision supermarkets, drugstores, and service stations will open up, just to fill their needs. And now, just suppose the radio doesn't sell. In a year we'll have to lay them all off. They'll be stuck for their monthly payments on homes and cars. Some of the stores and service stations will go bankrupt when the money stops rolling in. Their homes will go into sacrifice sales. Their kids, unless daddy finds a new job, will have to change schools. There will be a lot of heartaches all around, and that's not even thinking of my stockholders. And all this because you have made a design mistake. *That's* where your responsibility really lies, and I bet that they never taught you this at school!

I was very young and, frankly, impressed. Within the closed system of Mr. G.'s narrow market dialectics, it all made sense. Looking back at the scene from a vantage point of a good number of years, I must agree that the designer bears a responsibility for the way the products he designs are received at the market place. But this is still a narrow and parochial view. The designer's responsibility must go far beyond these considerations. His social and moral judgment must be brought into play long *before* he begins to design, since he has to make a judgment, an a priori judgment at that, as to whether the products he is asked to design or redesign merit his attention at all. In other words, will his design be on the side of the social good or not.

Food, shelter, and clothing: that is the way we have always described mankind's basic needs; with increasing sophistication we have added: tools and machines. But man has more basic needs than food, shelter, and clothing. We have taken clean air and pure water for granted for the first ten million years or so, but now this picture has changed drastically. While the reasons for our poisoned air and polluted streams and lakes are fairly complex, it must be admitted that the industrial designer and industry in general are certainly co-responsible with others for this appalling state of affairs.

In the mid-thirties the American image abroad was frequently created by the movies. The make-believe, fairyland, Cinderella-world of "Andy Hardy Goes To College" and "Scarface" communicated something which moved our foreign viewers more, directly and subliminally, than either plot or stars. It was the communication of an idealized environment, an environment upholstered and fitted out with all the latest gadgets available.

Today we export the products and gadgets themselves. And with the increasing cultural and technological Coca-colonization of that part of the world we are pleased to think of as "free," we are also in the business of exporting environments and "life styles" of the prevalent white, middle-class, middle-income society abroad and into ghettos, poverty pockets, Indian reservations, etc., at home.

The designer-planner is responsible for nearly all of our products and tools and nearly all of our environmental mistakes. He is responsible either through bad design or by default: by having thrown away his responsible creative abilities, by "not getting involved," or by "muddling through."

I am defining the social and moral responsibilities in design. For by repeating his mistakes a millionfold or more through designs affecting all of our environments, tools, machines, shelters, and transportation devices, the designer-planner has finally put murder onto a mass production basis.

Three diagrams will explain the lack of social engagement in design. If (in Diagram I) we equate the triangle with a *design problem*, we readily see that industry and its designers are concerned only with the tiny top portion, without addressing themselves to the real needs.

Let's take a rural mailbox for example. As it is now, it is usually large enough to hold letters and several magazines for a number of days. The structure is sheet metal and vaguely breadbox-shaped so that snow, ice, and rain will easily slide off. It also carries a small signal flag to be raised when the mail is delivered. It is inexpensive and sturdy.

Diagram 1: THE DESIGN PROBLEM

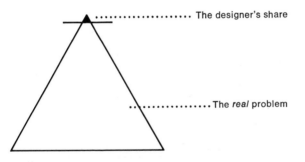

Quite recently a West Coast design office redesigned rural mailboxes for a national manufacturer. The result: a series of French Provincial, Japanese, colonial, or spaceship-inspired extravaganzas which are costly and which clutter up the visual landscape. They are "high-style" enough to be forced into obsolescence every few years and, incidentally, snow no longer slides off them. They will probably sell well in suburbia and exurbia and will take on some of the symbolic values of new status objects. The manufacturers are to be congratulated: many more mailboxes will be sold and, more importantly, many more can be pushed upon the public every few years as even fashion in mailboxes is manipulated.

What is the designer's evaluation? There is little wrong

Go ahead ...put your feet up!

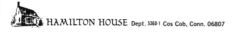

This idiot gadget is made and successfully sold in one part of the
world . . .

. . . while in another, this is a
family's sole means of cooking. Mexican stove from Jalisco, made of used
license plates and sold for about 8
cents. It is used as a charcoal brazier.
When the soldering finally pops after
some ten or fifteen years of use, it
is repaired, if possible, or else the
family has to invest another 8 cents
for a "new" stove. (Anonymous design, collected by John Frost, author's
collection.)

with the rural mailbox as it stands, apart from its cluttering
the landscape. But if redesign is called for, then the real
problems of rural mail delivery, in other words the huge
bottom area of our disgrammatic triangle, must be re-examined. To what extent can mailboxes be made to recede
(or even disappear) into the landscape? Can new materials,

tools, and processes reduce costs and, more importantly, reduce material waste? Can these containers be made tamper-
proof and vandal-proof? With heavier mail does the old size
still hold true? Can redesign help delivery? (With the incredibly antiquated mail delivery system in the United States
there is little doubt that electronic data-scanning procedures
will result in a more nearly normal delivery schedule: until
quite recently in England, for instance, city mail was delivered eight times daily, rural mail four times.) Should
the consumer in fact be required to buy mailboxes at all or
should a minimal Federal standard be written which eases
delivery procedures and guards privacy? Should local newspapers be permitted to add their own tubular "mailboxes,"
using them as shrill billboards which further befoul the edge
of the road? These are only a few of the questions which a
committed designer would ask himself; at last, most of the
triangle (top and bottom) would be explored.

Diagram 2: A COUNTRY

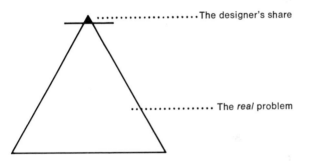

Diagram II is, of course, identical to Diagram I. Only the
labels have changed. For "Design Problem" we have substituted "Country." In a way, the justice of this becomes immediately apparent when talking about some far-off, exotic
place. If we let the entire triangle stand for nearly any
South or Central American nation (with the possible exception of the Republic of Cuba) we can see its telesic aptness.
Nearly all of these countries exist with wealth concentrated
in the hands of a small group of "absentee landlords." Many

of these people have never seen the South American country which they so efficiently "administer" and exploit. Design is a luxury enjoyed by a small clique who form the technological, moneyed, and cultural "elite" of each nation. The 90 per cent native Indian population which lives "up-country" has neither tools nor beds nor shelter nor schools nor hospitals that have ever been within breathing distance of designer's board or workbench. It is this huge population of the needy and the dispossessed who are represented by the bottom area of our triangle. If I suggest that this holds equally true of most of Africa, Southeast Asia, and the Middle East, there will be little disagreement.

Unfortunately, this diagram applies just as easily to our own country. The rural poor, the black and white citizens of our inner cities, the educational tools we use in over 90 per cent of our school systems, our hospitals, doctors' offices, diagnostic devices, farm tools, etc., suffer design neglect. New designs may sporadically occur in these areas, but usually only as a result of market pressures, rather than as a result of either research breakthrough or a genuine response to a real need. Here at home we too must assign those served by the designer to the minuscule upper part of the triangle.

Diagram 3:   THE WORLD

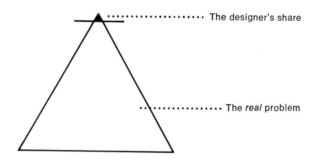

As the reader will have discovered by now, the third diagram is identical to the first and the second. But again we have changed labels. For now we call it "The World."

Can there be substantial doubt that the peoples of this world are not served by designers?

Where has our spirit of innovation gone? This is not an attempt to "take all the fun out of life." After all, it is only right and proper that "toys for adults" should be available to those willing to pay for them, and after all, as has been pointed out all too often, we live in an abundant society. But only a small part of our responsibility lies in the area of aesthetics. Sometimes one is tempted to ask why not *one* American table radio, for example, is well designed, whereas Sony, Hitachi, Panasonic, and Aiwa carry lines of some 84 highly specialized table radios, each one designed for a specific use area. (This record could easily be duplicated with tape recorders, TV sets, or, say, cameras.) After all, many book publishers, while pushing incredible trash onto best-seller lists, manage to bring out a few worthwhile volumes each year.

I am not necessarily pleading for extraordinary, innovative design for radios, alarm clocks, high-intensity lamps, refrigerators, or whatever; I am just hoping for product statements aesthetically acceptable enough not to conjure up visions of a breadbox raped by a Cadillac in heat. Isn't it too bad that so little design, so few products are really relevant to the needs of mankind? Watching the children of Biafra dying in living color while sipping a frost-beaded martini can be kicks for lots of people, but only until *their* town starts burning down. To an engaged designer, this way of life, this lack of design, is not acceptable.

All too often designers who try to operate within the entire triangle (problem country or world) find themselves accused of "designing for the minority." Apart from being foolish, this charge is completely false and reflects the misconception and misperception under which the design field operates. The nature of this faulty perception must be examined and cleared up.

Let us suppose that an industrial designer or an entire design office were to "specialize" exclusively within the areas of human needs outlined in this and other chapters. What would the work load consist of? There would be the design of teaching aids: teaching aids to be used in pre-nursery-school settings, nursery schools, kindergartens, primary and secondary schools, junior colleges, colleges and universities, graduate and post-doctoral research and study. There would be teaching aids and devices for such specialized fields as adult education, the teaching of both knowledge and skills to the retarded, the disadvantaged, and the handicapped; as well as special language studies, vocational re-education, the rehabilitation of prisoners, and mental defectives. Add to this the education in totally new skills for people about

Perch or reclining structure to be used in class-rooms *in addition* to regular chairs. This provides eight more positions for restless children. Designed by Steven Lynch, as a student at Purdue University.

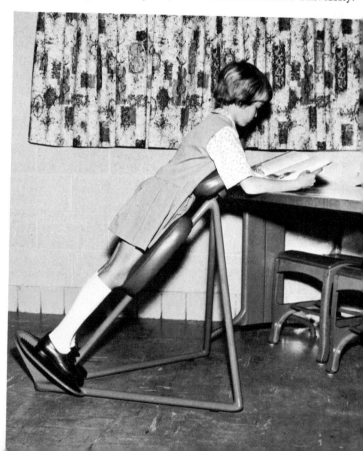

to undergo radical transformation in their habitats: from slum, ghetto, or rural poverty pocket to the city; from the milieu of, say, a central Australian aborigine to life in a technocratic society; from Earth to space or Mars; from the tranquility of the English countryside to life in the Mindanao Deep or the Arctic.

The design work done by our mythical office would include the design, invention, and development of medical diagnostic devices, hospital equipment, dental equipment, surgical tools and devices, equipment and furnishings for

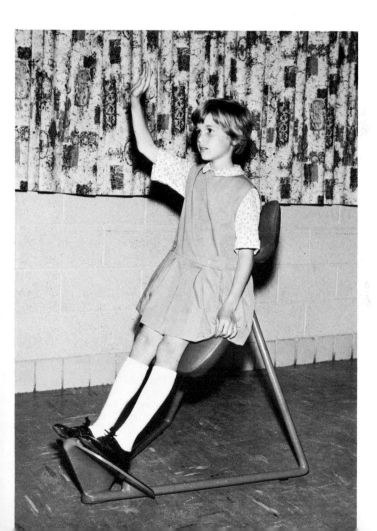

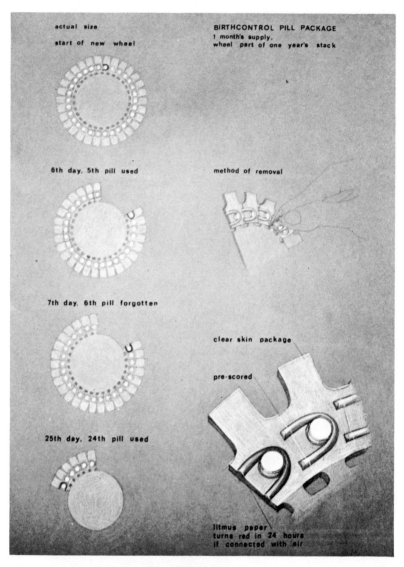

actual size

start of new wheel

BIRTHCONTROL PILL PACKAGE
1 month's supply,
wheel part of one year's stack

6th day, 5th pill used

method of removal

7th day, 6th pill forgotten

clear skin package

pre-scored

25th day, 24th pill used

litmus paper
turns red in 24 hours
if connected with air

ABOVE: Package for birth control pills for use by largely illiterate people. A run of placebos is included so that counting is unnecessary. If a user forgets to break off one day's pill from styrene wafer, the U-shaped tube turns red, as a reminder. Designed by Pirkko (Tintti) Sotamaa, Purdue University.

RIGHT: Canes for the blind, of hand-aligned fiber optics. They glow in the dark and also provide a more sensitive tactile feedback for the hand. Student designed by Robert Senn, Purdue University.

mental hospitals, obstetrician's equipment, diagnostic and training devices for ophthalmologists, etc. The range of things would go all the way from a better readout of a fever thermometer at home to such exotic devices as heart-lung machines, heart pacers, artificial organs, and cyborgian implants, and back again to humble visor-like eyeglasses, reading mechnisms for the blind, improved stethoscopes and urinalysis devices, hearing aids and improved calendrical dispensers for "the pill," etc.

The office would concern itself with safety devices for home, industry, transportation, and many other areas; and with pollution, both chemical and thermal, of rivers, streams, lakes, and oceans as well as air. The nearly 75 per cent of the world's people who live in poverty, starvation, and need would certainly occupy still more time in the already busy schedule of our theoretical office. But not only the underdeveloped and emergent countries of the world have special needs. These special needs abound at home as well. "Black lung" disease among the miners of Kentucky and West Virginia is just one of a myriad of occupational ills, many of which can be abolished through relevant re-design of equipment and/or processes.

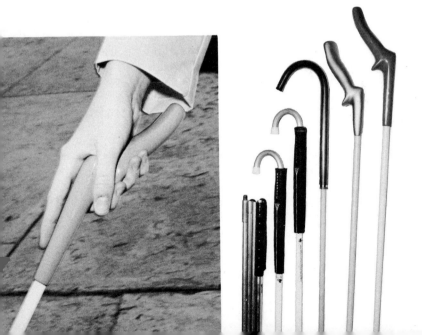

Middle and upper managerial ranks (if male and between ages of thirty-five and sixty) are a prime health-hazard group. The incidence of death from cardiovascular arrests by stroke or heart attack is frighteningly high. This loss of human lives can be ascribed to three main causes: faulty diet, a lack of exercise, and stress syndromes. Exercising equipment with built-in motivation might spare the lives of many people in this group, a group desperately needed all over the world to keep humanity going.

Basic shelters for American Indians and the Lapp population of Norway, Sweden, and Finland—and shelters (both temporary and permanent) for all men poised at the edge of an alien environment—need design and discovery. Whether it be a comparably simple shelter such as a space station or dome cities for Venus or Mars or something as complex as the complete "terraforming" of the moon, our design office will be needed here just as it is in sub-oceanic cities, Arctic factories, and artificial island cities to be anchored like so many pleasure boats in the Amazon River Basin, the Mediterranean, or around the (genuine) island chains of Japan and Indonesia.

Research tools are usually "stuck-together," "jury-rigged" contraptions, and advanced research is suffering from an absence of rationally designed equipment. From radar telescopes to simple chemical beakers, design has lagged behind. And what about the needs of the elderly and the senile? And of pregnant women and the obese? What about the alienation of young people all over the world? What about transportation (surely the fact that the American automobile is the most efficient killing device since the invention of the machine gun doesn't permit us to rest on our laurels), communication, and design for entirely new "breakthrough" concepts?

Are we still designing for minorities? The fact of the matter is that all of us are children at one point of our lives, and that we need education throughout our lives. Almost all of us become adolescent, middle-aged, and old. We all need the services and help of teachers, doctors, dentists, and hospitals. We all belong to special need groups, we all live in an underdeveloped and emergent country of the mind, no matter what our geographical or cultural location. We all need transportation, communication, products, tools, shelter, and clothing. We must have water and air that is clean. As a species we need the challenge of research, the promise of space, the fulfillment of knowledge.

If we then "lump together" all the seemingly little minorities of the last few pages, if we combine all these "special" needs, we find that we have designed for the majority after all. It is only the "industrial designer," style-happy in the seventies of this century, who, by concocting trivia for the market places of a few abundant societies, really designs for the minority.

Why this polemic? What is the answer? Not just for next year but for the future, and not just in one country but in the world. During the summer of 1968 I discovered a Finnish word dating back to medieval times. A word so obscure that many Finns have never even heard it. The word is: *kymmenykset*. It means the same thing as the medieval church word *tithe*. A tithe was something one paid: the peasant would set aside 10 per cent of his crop for the poor, the rich man would give up 10 per cent of his income at the end of the year to feed those in need. Being designers, we don't have to pay money in the form of *kymmenykset* or a tithe. Being designers, we can pay by giving 10 per cent of our crop of ideas and talents to the 75 per cent of mankind in need.

There will always be men like Buckminster Fuller who spend 100 per cent of their time designing for the needs of man. Most of the rest of us can't do that well, but I think that even the most successful designer can afford one tenth of his time for the needs of men. It is unimportant what the mechanics of the situation are: four hours out of every forty, one working day out of every ten, or ideally, every tenth year to be spent as a sort of sabbatical designing for many instead of designing for money.

Even if the corporate greed of many design offices makes this kind of design impossible, students should at least be encouraged to work in this manner. For in showing students new areas of engagement, we may set up alternate patterns of thinking about design problems. We may help them to develop the kind of social and moral responsibility that is needed in design.

Problems are everywhere. Left-handedness has never been designed for (see Chapter Six). The SDS used compelling rhetoric some years ago about "talking to the workers." But how about working with the workers? "Hard hats" are given their name because of the protective headgear they wear. But these hats are unsafe, not sufficiently tested for absorption of kinetic energy. Recently, when I had to buy several hard hats, I noticed that many of these come equipped with directions for cleaning, fitting them to the correct size, and much else. However, I should like to quote from the pamphlet of the "safety" helmet made by Jackson Products of Warren, Michigan:

> CAUTION: This helmet provides limited protection. It reduces the effect of the force of a falling object striking the top of the shell.
>
> Contact of this helmet shell with energized electrical conductors (live wires) or equipment, should be avoided. NEVER ALTER or MODIFY the shell or suspension system.
>
> Inspect regularly and replace suspension system and shell at first sign of wear or damage.
>
> *The WARNING stated above applies to all industrial safety hats and caps, regardless of manufacturer.* (My italics.)

This last statement really seems to be true since all hard-hat pamphlets carry this warning, using almost identical words.

The nearly two million safety goggles manufactured annually in this country are unsafe—the lenses scratch easily, some may shatter, and most crack the bridge of the nose under a blow. So-called "hard shoes," designed to protect the front of the foot against falling debris, do not absorb sufficient kinetic energy to be useful; the steep cap over the toes can be crushed by a small steel beam falling one yard. Most cabs in long-distance trucks vibrate so that they will materi-

ally destroy a man's kidneys in 4 to 10 years. The list could
go on.

What may be needed here is a designers' commune. Most
communes in this country have determinedly marched into
the past. But baking bread, playing a guitar, weaving
fabrics, and doing ceramics are not the only rational alternatives to a consumer society. Nor is the mind-blowing violence of a Charles Manson. With most of the communes
poised in a choice between nihilism and nostalgia, a commune of planners and designers might prove to be the best
alternative. (I've written more about this in Chapter Twelve.)

There must have been a time a few million years ago
when some nameless early caveman killed a rabbit, ate it
in his cave, and threw the bones on the ground. And surely
his wife implored him to throw the bones out of the cave, to
keep it neat and clean. Times have changed. We are all in
the same cave together, and there is no longer any place to
throw the trash. Or to change to a more meaningful meta-
phor, we are all together on this small spaceship called
"Earth," 9,700 miles in diameter and sailing through the
vast oceans of space. It's a small spaceship and 50 to 60
per cent of the population cannot help to run it, or even help
themselves stay alive, through no fault of their own. Where
hunger and poverty lead small children to eat the paint off
walls and die of lead poisoning in Chicago and New York
ghettos. Where children in Los Angeles and Boston die of
infected rat bites. To deprive ourselves of the brain and
potential of any person on our spaceship is wrong and no
longer acceptable.

The haves seldom realize that during the past decade
the difference between haves and have-nots has become
greater than ever before in history, and that this difference
has grown even more frighteningly in the last year than
during the preceding nine years. The chasm that divides
those who live with the highest power ratio and have ample
food and those who don't is widening all the time, and
there are more and more people in the have-not lands. The
island of Mauritius had a population of 226,000 in 1941;
deaths accounted for 36 out of every 1,000 people. By 1955,

648,000 people lived on Mauritius *without any gains through immigration*. Through public health services the death rate had been reduced to 12 in 1,000. This health picture will change even more drastically. We can estimate that through an increased birth-rate and a cut in infant mortality, this small island will contain more than one million people by the year 2000. People who will need to live, eat, and consume certain minimum goods. Mauritius has little agriculture and no industry to speak of. By using this one island as an example, we can see some of the real needs (hidden for a million years) beginning to emerge for the first time.

All this raises the question of value. If we have seen that the designer is powerful enough (by affecting all of man's tools and environment) to put murder on a mass production basis, we have also seen that this imposes great moral and social responsibilities. I have tried to demonstrate that by freely giving 10 per cent of his time, talents, and skills the designer can help. But help where? What is a need?

In the early fifties I had the good fortune to enjoy a lengthy correspondence with the late Dr. Robert Lindner of Baltimore. Together we worked on a book to be called *Creativity Versus Conformity*, a collaboration ending only with his untimely death. I should like to quote extensively from the Prologue (pp. 3–6) of his *Prescription for Rebellion* concerning his concept of value:

> The end to which man studies himself cannot be other than to realize the full potentiality of his being, and to conquer the *triad of limitations* fate or God, or destiny, or sheer accident, has imposed on him. Human beings are enclosed by an iron triangle that forms for their race a veritable prison cell. One side of this triangle is the medium in which they must live; the second is the equipment they have, or can fashion, with which to live; the third is the fact of their mortality. All effort, all being, is directed upon the elimination of the sides of this enclosure. If there is purpose to life, that purpose must be to break through the triangle that thus imprisons humanity into a new order of existence where such a triad of limitations no longer obtains. This is the end toward which both individual and species function; this is the

end toward which the race strives; this is the end which gives meaning and substance to life.

Behind and beyond the word-games philosophers play, and in the final analysis, all that man does—alone or in the organizations he erects—has as its design the overcoming of one or more or all aspects of this basic triad of limitations. What we call progress is nothing more than the small victories every man or every age wins over any or all of the sides of the imprisoning triangle. Thus progress, in this one and only possible sense, is a measurable thing against which the sole existence of a person, the activities and aims of a group, even the achievements of a culture, can be estimated and assigned value.

The as yet uncalculated millennia during which man has tenanted Earth have been witness to his continued valiant efforts to escape from the triangle that interns him. Inexorably and against odds, over the centuries, he has fought against and conquered the medium of his habitat until now he stands poised on a springboard to the stars. Today, earthbound no longer, and loosening even the fetters of gravity, he can look backward to count his conquests. The elements have succumbed to him, and also the natural barriers of space and time. Once confined to a small area bounded by the height of the trees he could climb, the distance his legs could carry him, the view his eye could encompass, the length his voice could carry, the reach of his arms, and the acuity of his remaining senses—once a cowed victim of every hazard to existence vagrant Nature has in her catalogue—now he is lord over those containing powers that would have held him slave to them forever. So one iron wall of his cell has been worn thin, and, through the vents and cracks he has made in it, come far-traveled winds of freedom and the beckoning gleam of the universes outside.

Similarly, the second side of the triangle—the limitations imposed by the biologically given equipment of human beings—has yielded step by step to the ongoing, persistent struggle against it by men. In the main, this has been a process of extension. It has been marked by the fashioning of tools to improve the uses of the limbs, the sensitivities of the specialized end-organs and the efficiency of those other parts and organs that complete the body. Here the victories have been of an immense order of magnitude. They have culminated in what amounts to a total breakthrough of the envelope of skin that enwraps us, even to the point where the products of hand and brain—as in the giant computing machines and other physical miracles of our time—by far outdo many capacities of their creators. And, finally, in the matter of the last side of the triangle, while the days of our years still last but an eyewink on the bland face of the eternal clock, longevity if not immortality is now more than a promise.

The uses of knowledge are clear despite the turgid morass through which a seeker must plow to find order and sense therein. The sciences and arts—like the individual lives men live—are all strivings and experiments. They are pointed toward the realization of human potentiality and ultimately contributory to that evolutionary breakthrough which will come when the walls of the containing triangle finally crash to earth. Thus the value of an item of knowledge, an entire discipline, or a deed of art can be placed upon a scale, and its measure also taken.

Much as we have established a six-sided "function complex" in order to evaluate design in the first chapter, we can now plug in the "triad of limitations" and use it as a primary filter to establish the social value of the design act. While the American automobile is examined in great detail in a later chapter, it can be used as a demonstration object now.

Early automobiles overcame one of the three prison walls of the triad. It was possible to go farther and faster in an automobile than a human being's legs would carry him, and to carry a heavy load as well. But today the automobile has become so overloaded with false values that it has emerged as a full-blown status symbol, dangerous rather than convenient. It breathes and exhales a great amount of cancer-inducing fumes, it is overly fast, wastes raw materials, is clumsy, and kills 50,000 people in an average year. On an average weekday the time needed in rush hour to go from the East River to the Hudson on Forty-second Street in New York is at least one hour: a man walking can easily do it in but a fraction of that time. Considering these aspects, the concept of the automobile has been manipulated so that it now shores up the wall of mortality in the triad; its contributions have grown negligible by comparison.

The car, however, is only one example. Everything designed by man can be forced through the filter and evaluated in a similar matter.

K. G. Pontus Hultén's *The Machine as Seen at the End of the Mechanical Age* (1968) is an excellent book. Two quotes from it are relevant here. Commenting on Jean Tinguely's "Rotozaza No. 1," Hultén says:

> The production of articles that nobody really needs, but which occupy the ground floors of all big stores, is one of the many outward symptoms of something basically wrong in a world of overproduction and undernourishment. In order to control overproduction, without going through the intricacies of selling the product, it becomes necessary for a wilfully destructive war to be going on permanently somewhere. Today, the world is spending over $150 billion per annum on the actual or potential destruction of lives and property, as compared with the capital transfer from rich to poor countries of about $10 billion per year—including a large share for military aid.

And on the following page, in reference to another work, he continues:

> Probably the greatest political problem facing the world today is the difference among various regions as regards their technological development. Many parts of Europe and America are already leaving the mechanical age to enter the electronic era, while much of Africa, for example, is only beginning to be industrialized.
>
> To some extent, the mechanical age seems linked to the age of colonialism. Both reached their apogee in the nineteenth century; both were based on the instinct for exploitation. The world was prospected to discover and cultivate raw materials with which to feed the machines. It rarely occurred to the ruling powers that the people whose soil produced these materials, and who sweated to bring them forth, should have any appreciable use and benefit from other products. Whenever the natives made any serious trouble, the usual response was to send a gunboat.
>
> Up to 1950, there were four independent countries in

Africa; today, there are more than forty. They are politically aware and highly nationalistic, but technologically extremely underdeveloped. Industrial output in all of Africa (except South Africa) is, in fact, less than that in Sweden alone. Unless foreign governments and private corporations unite with the African nations in a massive and long-range program of industrial development, the social and political results will probably be explosive.

Let me elaborate on this with some very concrete metaphors from the market place (which Americans might do well to recite nightly before retiring): The retail price of *one* Aston Martin DB is equal to that of 14 emergency public health clinics in the Third World. The research monies tied up in developing the Ford "Maverick" (just one of many models) would provide technical schools for *all* of Africa. In Japan, the camera industry has recently unveiled the Nikon Photomic "FTN" with an automatic exposure meter. The price for this camera (with a normal range of accessories) is equal to that of 8 fishing boats in the South Pacific. Finally, the Nikon "FTN" is largely sold to amateur photographers who solemnly replace last year's Nikon "F" (without the meter) with it. Both cameras dramatically exceed most normal requirements of amateur photography, especially with the introduction of new, high-speed color films. Of course, all of us are well aware that the American housewife spends far more annually on beauty shops and cosmetics than the nation spends for education. And perhaps somewhat less aware that it costs $50,000 to $250,000 to kill one Vietnamese.

What needs to be done? And how can we do it? A series of examples may serve as the best answer.

One of the world's few really great designs for emergent countries was developed during the last 25 years by a team of 3 designers from as many different countries. It is a brick-making machine. This simple device is used as follows: Mud or earth is packed into a brick-shaped receptacle, a large lever is pulled down, and a perfect "rammed earth" brick results. This apparatus permits people to "manufacture" bricks at their own speed—500,000 a day or 2 a week. Out

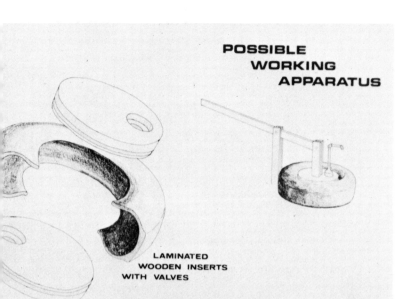

POSSIBLE
WORKING
APPARATUS

LAMINATED
WOODEN INSERTS
WITH VALVES

PUMPING DEVICE

FLUID
FLOW-
EXIT

APPLIED
FORCE

OPERATION

FLUID FLOW-INTAKE

One of a series of twenty investigations into the use of old tires, which now abound in the Third World. Both of these irrigation pumps have since been built and verified. Designed by Robert Toering, as a student at Purdue University.

of these bricks schools, homes, and hospitals have been built all over South America and the rest of the Third World. Today schools, hospitals, and entire villages stand in Ecuador, Venezuela, Ghana, Nigeria, Tanzania, and many other parts of the world. The concept is a great one: it has kept the rain off the heads of people, and it has made instruction possible in schools where it was not possible before and where the schools themselves did not exist a few years ago. The brick machine has made it possible to construct factories and install equipment in areas where this had never been attempted in the past. This is socially conscious design, relevant to the needs of people in the world today.

During the international design festival at Jyväskylä, Finland, in 1968, I participated as part of a UNESCO team of international design experts to develop new ideas for Black Africa. Many problems await solution. The circulatory system of the Third World, and specifically Black Africa, is

Low-cost educational TV set to be built by Africans in Africa. Designed by Richard Powers, as a student at Purdue University.

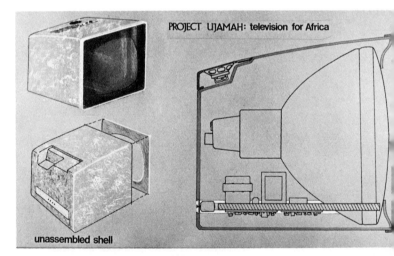

RIGHT: Another variation of African TV proposal. Designed by Michael Crotty, as a student at Purdue University.

very bad. The people get sick because waste products cannot be efficiently rinsed away; there is almost no sanitation. There is not enough water because water is polluted by precipitation, by flowing through open ditches, and by incredibly fast evaporation. Often water is uncontrolled and washes away precious topsoil. Irrigation is virtually non-existent in the villages. The missing element is a *pipe,* or rather a simple device that will make it possible to manufacture pipe segments in the village, by "cottage industry" or by an individual. So the task is to design a pipe-making machine. A pipe-making machine that can be built in Africa by Africans and used for the common good. A machine (or tool) that will by-pass private profit, corporate structures, exploitation, and neo-colonialism.

Black men from seven nations told me that one of their greatest needs was an inexpensive educational TV set. This will be a set to be distributed to African states through UNESCO, to be made in Africa, using native materials as far as possible, as well as local labor. It should give no

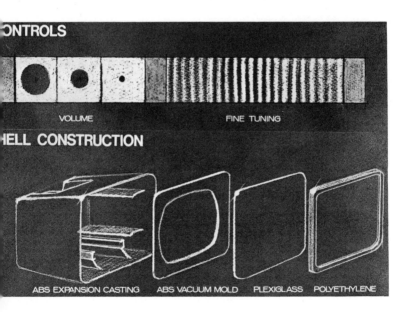

68 DESIGN FOR THE REAL WORLD

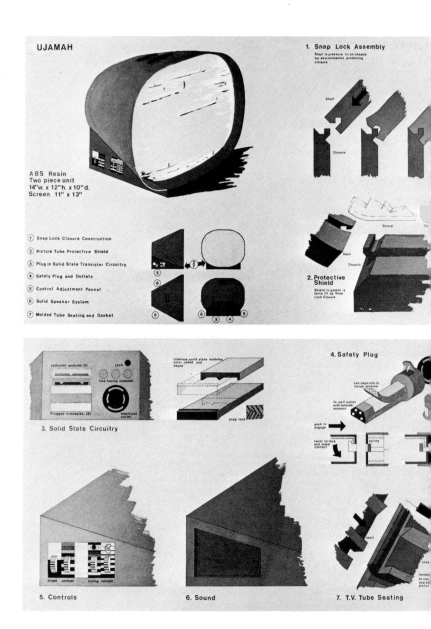

profit to any private corporation in Europe or North America.

Television was developed in Great Britain and the United States nearly 40 years ago. Since these were the first countries to develop it and because of their market structure, set design has been frozen into early levels, technologically. TV sets in North America show images with a line resolution of 525 lines to the inch. Russian sets have 625; Great Britain has 405 and 625. French sets have a line resolution of 819 to the inch. This means that these images are clearer and demand less from the eye and brain to decode the information. Obviously a television set that is completely new, devoted primarily to educational needs, should carry a high line resolution.

In doing the basic research for this set, my graduate students and I have found much to astonish and delight us. Even in a sophisticated, technologically advanced country like Germany, TV sets have selectors for 13 different channels, even though only 2 channels are used. In our

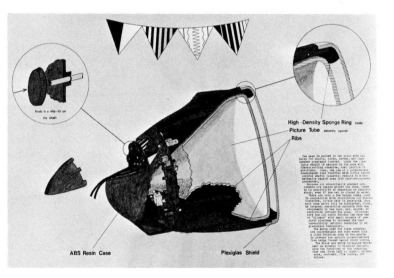

Knob is a slip-fit on the shaft

High-Density Sponge Ring holds Picture Tube securely against Ribs

ABS Resin Case

Plexiglas Shield

case this entire selection mechanism can safely be left out because we are working on a one-channel set. The eventual breakdown of the vacuum picture tube is speeded up through the use of on-off switches. We plan to broadcast educational material only, and the set will always be "on." Current drain is negligible with the use of transistors. Since the set must be tropicalized, venting it seemed to be crucial. Tropicalization in Africa, however, also means preventing the entry of thousands of small insects. We found that through the use of integrated circuits, internal heat build-up was negligible enough to eliminate all need for fans and/or vents. In fact, by burying an aluminum heat sink in the housing of the set, enough heat is drained off.

Our research had to consider climatology, anthropology, electrics and electronics, population densities, prevalence of African languages in various areas, terrain (for transmission reasons), social attitudes, and many other guidelines of design.

After we discovered, somewhat late in our research, that a highly sophisticated, market-competitive TV set (including 36 channels, on-off switching, internal fans, and an inordinate amount of "sexy" styling) sold for $119.95 retail in the United States (including all profits, shipping, and customs charges), we investigated that set. We found that labor, manufacturing costs, and material cost the Japanese manufacturer less than $18. We have accepted the fact that in terms of existing African skills, factories, and distribution networks, our set can be made for considerably less than $9.

Our final concept of the shell of the TV set will permit both mass production (at the rate of 2,000 shells per machine-day) or individual production by one individual standing in a village street. It can then be *his* decision as to whether he wishes to make 30 shells a day or one every few days.

At the time of this writing (December 1970), the introduction of video-tape cartridges in both black and white and color is only a matter of time. It is for this reason that we are redesigning our educational television set so that it will also accommodate the cartridge. There can be no doubt that, especially in areas of education, video cartridges will completely revolutionize the development of the Third World.

In a short time this $9 TV set will be completed and given to UNESCO. It will join our 9¢ non-electric, thermocoupled, cow-dung-powered radio (designed for Indonesia).

There are many ways of working for the needs of underdeveloped and emergent countries. The simplest, most often employed, and probably shabbiest is for the designer to sit in his New York, London, or Stockholm office and to design things to be made in, say, Tanzania. Souvenir-like objects are then manufactured, using native materials and skills, with the pious hope that they will sell in developed countries. They do, but for a short while only, for by designing "decorative objects for the home" and "fashion accessories," we merely tie the economy of that country to the economy of other countries. Only two possibilities remain: Should the economy of the wealthy Western country fail, the emergent country's new economic independence fails with it. Should the economy of the wealthy Western country continue climbing, the fashion likes and dislikes of its population will be manipulated even more, and consequently, the emergent country's new economic independence will fail also.

A second, slightly more effective way for the designer to participate would be to spend some time in the underdeveloped country developing designs really suited to the needs of the people there. This still begs the question of meaningful engagement.

A somewhat better way: to move the designer to the underdeveloped country and have him train designers there, as well as designing and working out the logistics of design needs for that nation. But even this is no ideal solution.

Ideally (as things stand now): The designer would move to the country and do all the things indicated above. But in addition, he would also train designers to train designers. In other words he would become a "seed project" helping to form a corps of able designers out of the indigenous population of the country. Thus within one generation at most, five years at the least, he would be able to create a group of designers firmly committed to their own cultural heritage, their own life style, and their own needs.

All design must be *operative*. Just as the African TV set becomes synergetically more than a TV set as its manufacture requires new native skills, a large pool of labor reserve, new factories, and new communication channels, design itself must, as shown in the above example, always be a seed project, always operative.

5

OUR KLEENEX

CULTURE:

Obsolescence,

Permanence, and Value

*You have to make up your mind
either to make sense or to make
money, if you want to be a designer.*
R. BUCKMINSTER FULLER

IN ALL LIKELIHOOD it started with automobiles. Dies, tools, and molds that are used in manufacturing cars wear out after about 3 years of usage. This has provided the Detroit automobile makers with a timetable for their "styling cycle." Minor cosmetic changes are performed at least once a year; because of the need for rebuilt and redesigned dies, a major style change is plugged in every 2½ or 3 years. Since the end of World War II, the car manufacturers have sold the American public on the concept that it is stylish and "in" to change cars every 3 years, at the very least. With this continuous change has come sloppy workmanship and virtually non-existent quality control. For a quarter of a century American national administrations have proclaimed their tacit approval or enthusiastic support of this system. Some of the economic and waste-making results of this policy have been documented in other chapters. But what is at stake here is an expansion: from changing automobiles every few years, to considering everything a throw-away item, and considering *all* consumer goods, and indeed, most human values, to be disposable.

When people are persuaded, advertised, propagandized, and victimized into throwing away their cars every 3 years, their clothes twice yearly, their high-fidelity sets every few years, their houses every 5 years (the average American family was moving once every 56 months as I was writing this book), then we may consider most other things fully obsolete. Throwing away furniture, transportation, clothing, and appliances may soon lead us to feel that marriages (and other personal relationships) are throw-away items as well and that on a global scale countries, and indeed entire sub-continents, are disposable like Kleenex.

In spite of its shrill and horrifying overtones, Vance Packard's *The Wastemakers* tells the story of forced obsolescence and the death-rating of products like it is.

That which we throw away, we fail to value. When we design and plan things to be discarded, we exercise insufficient care in designing or in considering safety factors.

On April 8, 1969, the Health Department of Suffolk County, New York, reported on their study of color TV sets. Studying sets of all sizes, prices and makes, they discovered that minimally 20 per cent of all color TV sets emitted harmful X-rays at a distance of from 3 to 9 feet. In other words, at least one in every 5 color television sets being used may sterilize the viewer, or subject his or her children to genetic damage after prolonged exposure.

As of April 1, 1969, General Motors was recalling one out of every 7 automobiles and trucks for "remedial repairs" as these vehicles proved themselves clearly unsafe in operation.

More than $750,000 has been awarded to plaintiffs or *their survivors* by a jury in Sacramento because of the negligent design of the gas tank on the early Corvette. A jury in Los Angeles gave more than $1 million to several people because of the Volkswagen's poor cornering capability.

On June 29, 1967, Ernest Pelton, a seventeen-year-old

playing football for his high school in Sacramento County, California, received a head injury. The depression of the subcortical layers of his brain have plunged him into a permanent coma, and he is not expected ever to regain consciousness. Medical costs for the remainder of his lifetime have been estimated to run in excess of $1 million. What makes the story relevant is that Mr. Pelton wore the best and most expensive ($28.95) football helmet being manufactured. Every year 125,000 of these helmets are sold, yet *they have never been tested for absorption of kinetic energy!* In fact, of the 15 million safety helmets, hard hats, football helmets, etc., sold annually in this country, none have ever been tested in kinetic energy situations!

And the examples, culled from various news accounts over recent years, go on:

A young man is paralyzed for life because the power on his workbench is accidentally activated;

A mother of 3 is killed because her chest is crushed by the steering column in a car crash;

The boom of a large construction crane collapses, and 5 families are left without husbands or fathers;

Six hundred women (a year) lose their hands in top-loading washers;

A young girl leaving a drugstore is literally cut into ribbons because the plate glass door fails to pivot properly after a pebble gets caught in its track;

A crooked union boss fails to see that safety laws on cableways are enforced: a cable shears and 3 miners are crushed;

A tank car carrying carbon dioxide at $-20°$ C. explodes in the middle of a Midwestern town;

Three children are paralyzed from the neck down because they went down a slide head first (slides are badly designed, yet no attention has been given to redesigning them);

A gymnast is made into a quadriplegic because his portable horizontal bar is inherently unstable;

A baby drinks a toxic household cleaner and is brain-damaged for life.

It is probably impossible to make even a ballpark guess

regarding the number of deaths and injuries caused by design. There are a few figures we can start with, however. According to the National Safety Council, we kill an average of 50,000 Americans annually and maim a further 600,000 each year in traffic accidents. According to a report by Dennis Bracken (radio station KNX, Los Angeles, November, 1970), we kill and injure 700,000 children through unsafe toys each year in the United States. According to the National Heart Association, the lives of approximately 50 per cent of all industrial workers are shortened by 5 years or more through heart stress caused by noisy equipment. Unsafe home appliances account for 250,000 injuries and deaths annually. Even the design of so-called "safety equipment" imposes further and greater hazards: "Approved" fire escapes tend to fry people trying to use them. Eight thousand people have died this way over the years when they were trapped by the escape mechanism.

Recently the control panel on stoves has been moved to the rear of both gas and electric units. The manufacturer's explanation is that this will make control knobs more difficult for small children to reach. In reality, a merchandising gimmick is at work here: by running wiring straight up the back of the stove, the stove can be built less expensively, yet sold for more money. The control unit is still there, an attractive nuisance, and children merely have to climb upon a stool and balance precariously while trying to play with the pretty knobs. Often they may fall and burn their arms or faces. A design solution would be simple: a double-security switch that requires both hands to engage "on" (similar to "Record" switches on tape recorders). Instead, appliance manufacturers woo the public with such felicitous confections as a recent Hotpoint range, the oven of which played "Tenderly" whenever the roast was ready. (!)

(Since I collect examples of the idiocies dreamt up by my fellow designers, I found myself enchanted by two new offerings for the 1970 Christmas gift market. One of these was an electric, dial-a-matic Necktie Selector for the home. You push a series of buttons, specifying shirt color, suit color, and other pertinent data, then a little wheel moves forward and presents the 6 or 10 ties that fit in with your particular color choice. This gadget mounts on the inside of the closet door, comes in a choice of "modernistic" or "Early American," and is only $49.95. The other, alas, was still in the developmental stages in 1970, but they were promising it by *next* Christmas. It was an *electronic* necktie selector that uses a colorimeter and a scanning device to assess your entire wardrobe. No longer will you have to feed color specifications to the gadget and push buttons: instead, the tie selector will take a good hard look at you, scan your "ensemble," and then hand you the tie that is good for you. It seems we will be privileged to buy this for a mere 300 bucks a throw.)

There is no question that the concept of obsolescence can be a sound one. Disposable hospital syringes, for instance, eliminate some of the need for costly autoclaves and other sterilizing equipment. In underdeveloped countries, or climatic situations where sterilization becomes difficult or impossible, a whole line of disposable surgical and dental instruments will become useful. Throw-away Kleenex, diapers, etc., are certainly welcome.

But when a new category of objects is designed for disposability, two new parameters must enter the design process. For one thing, does the price of the object reflect its ephemeral character? The 99¢ paper dresses cited before are excellent answers to changing fashion or to travel needs, or in the area of temporary protective clothing. But this is not the case with the $149.50 paper dress.

The second consideration deals with what happens to the disposable article after it has been disposed of. Automobile junkyards follow our highways from coast to coast. And even these appalling smears on the landscape at least have a (painfully slow) rusting process in their favor, so that five or twenty years hence they will have turned to dust. The new plastics and aluminum will not disintegrate, and the concept of being up to our armpits in discarded beer cans is not a pleasant prospect for the future.

It is here that bio-degradable materials (i.e., plastics that become absorbed into the soil, water run-off, or air) will have to be used more and more in the future. The Tetra-Pak Company, responsible for the distribution of seven billion milk, cream, and other packages a year, is now working on an ideal self-destructive package in Sweden. A new process developed in collaboration with the Institute for Polymer Technology in Stockholm, accelerates the decomposition rate of polyethylene plastics. Thus, packages will decompose much more rapidly after they have been discarded without having their strength and other properties affected while still in use. A new disposable, self-destructive beer bottle called "Rigello" is already on the market. Much more than these Swedish experiments will have to be done to save us from product pollution.

Fortunately, it has now become possible to use *the actual process of pollution* to bring about positive results. The result of a recent design research problem conducted with two graduate students is a good example.

We began by studying cockleburs, burdocks, and other botanical seeds that possess "hooking mechanisms." Out of this we developed an artificial hooking seed, approximately 40 cm. in length and made out of a bio-degradable plastic. The particular plastic chosen has a half-life of about 6 to 8 years. All plastic surfaces of these constructs are dipped in plant seeds and encapsulated in a hydrotropic nutrient solution. These "macro-seeds" are furnished folded flat, but spring-loaded, 144 to a package. When the package rips open, the macro-seeds spring into shape and (assuming there are hundreds of them) become inextricably hooked to one another (see illustrations). The theoretical concept is an extraordinarily simple one. It is possible to drop thousands of these seeds from airplanes into dry-wash areas of arid, desert-like country. Once dropped, the seeds spring open and interlink. With the first rain or even a substantial increase in moisture content of the air, the plant seeds on the surface of the artificial seeds begin to sprout (helped along by the nutrient solution encapsulating them). The macro-seeds themselves, helped along by these newly sprouted organic seedlings, now form a low but continuous

Artificial burrs, 40cm long, made of bio-degradable plastic and coated with plant seeds and a growth-boosting solution. To reverse erosion cycles in arid regions. Designed by James Herold and Jolan Truan, as students at Purdue University.

dam. (Such a dam can be theoretically infinite in length, and would be around 20 to 30 cm. in height. The experimental dam we constructed at a dry-wash area is 17 meters long.)

The dam, consisting by now of macro-seeds that are hooked together and further augmented by true organic growth, begins to catch the first spring run-offs. Seeds, mulch, topsoil, and other organic principles are captured by it; the dam grows both literally and figuratively. Within 3 to 6 seasons it has grown into a compact area of vegetation and a permanent trap for capturing topsoil. Towards the end of this time period the bio-degradable plastic core begins to be absorbed by the surrounding vegetation and soil and turns into a fertilizing agent.

Experimentally at least, the erosion cycle has been halted and in fact reversed.

The component factors of obsolescence, disposability, and self-destruction have been used for an ecological alteration, and an attempt has been made to salvage desert areas by introducing new thinking into design and planning.

Other applications of this—such as "throw-away" maple-seed-shaped extinguishers for forest fires—are discussed in greater detail in Chapters Seven and Nine.

To return to the primary concept of a disposable society: With increasing technological obsolescence, the exchange of products for newer, radically improved versions makes sense. Unfortunately, as yet there has been no reaction to this new factor on the market level. If we are to "trade in" yesterday's products and appliances for today's, and today's for tomorrow's at an ever-accelerating rate, then unit cost must reflect this tendency. Slowly, two methods of dealing with this problem are beginning to emerge.

Leasing rather than owning is beginning to make headway. There are a number of states in which it is less expensive to lease an automobile on a 3-year contract than to own one. This concept has the added motivation built in that the man who leases his automobile is no longer bothered by maintenance cost, insurance, fluctuating trade-n values, etc. In some of our larger cities it has become possible to lease such large appliances as refrigerators, freezers, stoves, dishwashers, washer and dryers, air-conditioning units, and TV sets. This trend has grown even more pronounced in manufacturing and office situations. Maintenance and service problems surrounding the hardware in the computer, research lab, and office-filing fields make the leasing of equipment more and more rational. Property tax laws in many states are also helping to make the concept of "temporary use" rather than "permanent ownership" more palatable to the consuming public.

It now becomes necessary only to convince the consumer that, in point of fact, he *owns* very little even now. The homes which make up our suburbs and exurbs are purchased on 20- or 30-year mortgages, but (as we have seen above) with the average family moving every 56 months, are sold and resold many times over. Most automobiles are purchased on an installment plan lasting 36 months. They are usually traded in 4 to 6 months before the contract is completed, and the still partially unpaid-for car is used as a trade-in. The concept of ownership, as it applies to cars, homes, and large appliances in a highly mobile society, becomes a mere polite fiction.

This is indeed a major volte-face regarding possessions. It is a change of attitude often condemned out of hand by the older generation (who are sublimely unaware of how little they themselves, in fact, ever own). But this moral condemnation is not really relevant and never has been. The "curse of possessions" has been viewed with alarm by religious leaders, philosophers, and social thinkers throughout human history. And the concept of being owned by things, rather than owning them, is becoming clear to our young people. Our greatest hope in turning away from a gadget-happy, goods-oriented, consumption-motivated society based on private capitalist acquisitive philosophies, lies in a recognition of these facts.

A second way of dealing with the technological obsolescing of products lies in restructuring prices for the consumer market. On Sunday, April 6, 1969, *The New York Times* carried an advertisement for an inflatable easy chair (imported from England) at a retail price of $6.95 (including shipping, taxes and import duties). Within 5 days mail and telephone orders were received for 60,000 chairs. A few years ago hassocks and occasional chairs made of plastic-reinforced cardboard were available at such discount outlets as Pier 1 and Cost Plus at prices ranging from 59¢ to $1.49. Such items, combining usefulness, bright color, modish design, comfort, extremely low cost, light weight and easy "knock-down factors" with eventual nonchalant disposability, naturally appeal to young people and college students. But their appeal is filtering down to larger and more "settled" segments of the population as well.

Mass production and automation procedures should make an increasing number of inexpensive, semi-disposable products available to the public. If this trend continues (*and does not lead to waste-making and pollution*), it is a healthy one. If further justification were needed for this throw-away society on moral grounds, a corollary trend has already begun to make its appearance. A home containing inexpensive plastic dinnerware will, often as not, also contain one or two pieces of fine craftsman-produced ceramics. The 99¢ paper dress will be dramatized with a custom-designed, custom-made ring created specifically for the wearer by a silversmith. One of the inexpensive cardboard easy chairs (bought a few years ago at Cost Plus) may well contain a $60 hand-woven cushion (bought at a prestigious craft shop or a gallery). In fact, the current renaissance of the crafts is partially traceable to consumer money liberated for investment in these custom-made art objects through the reduction of prices for everyday goods.

This trend is by no means in full force as yet. But if we look to moving day in 1999, we may well see a family loading their car with a few boxes full of art and craft objects, ceramics, hand-woven cushions and wall-hangings, books and cassettes while nearly all of the so-called "hard goods" have either been returned to the lessor or thrown away, to be replaced by newly leased appliances and inexpensively purchased furnishings at the point of destination.

There is one more factor that should be considered. Industrial design has become polarized more and more into shroud design, or "styling," and integrated design. Don Wallance, in his book *Shaping America's Products*, makes the useful distinction between objects whose inner structure and outer form are integral—such as a pottery bowl or a plywood chair—and objects where the outer form merely sheathes or shrouds an inner structure or mechanism—objects like refrigerators, radios, or locomotives, whose "essential nature separates the process of making from the process of designing." This distinction can of course remain a useful tool only as long as we stay comfortably in the middle of these classifications, without encountering overlap areas where such differences might become blurred.

When considering an object where both inner structure and exterior form are integral (such as a hunting knife, a teacup, a wine glass, or a pair of shears) it seems that we all, designers and consumers alike, can agree upon what constitutes good design. Without too much difficulty we can feed the design through our six-part function complex (cf.: Chapter One) and determine whether it is organic and functional.

When dealing with objects in the other categories, however, where the outer form merely serves as a shroud of an inner construct which designer and public agree should be covered up (such as an eighteenth-century grandfather clock, a present-day toaster, or a TV set), we find ourselves beset by doubts regarding the appropriateness of the design. The whole question is quite complex and has, in fact, become a question of personal taste. To quote briefly from John A. Kouwenhoven's *The Beer Can by the Highway:*

> . . . we may as well define taste as that sort of preference for one or another form which is relevant only when form is independent of function. Or, to put it differently, taste is that sort of form-preference which can logically be illogical, and usually is. Taste can have little to do with the design of an airplane wing or a propellor blade, but almost everything to do with the design of a refrigerator cabinet or a woman's dress.

Frequently the argument has been made that shroud design is justifiable as an aesthetic necessity, to cover up complex and messy internal arrangements of wiring, ducts, screws, rivets, etc., which would prove confusing to the viewer. However this argument is false. For with an airplane, racing car, or three-point hydrofoil boat, the *total Gestalt* often *is* the object.

John Kouwenhoven discusses the two rather different meanings one could apply to the word functionalism. One of these is *integral design*, which has much to do with the appropriateness of design to tools, material, and use and tends towards simplicity and what Horatio Greenough called "the majesty of the essential." *Sheath design*, or "shrouding," is the other kind of functionalism, and it seems to have less to do with the structure of the object than with the structure of the designer's or the consumers' psyches. This may whimsically move towards the baroque or the simple. Kouwenhoven suggests that we call this kind of functionalism "effective packaging," and let it go at that. Certainly the entire concept of automotive styling seems to fit into this category.

These types of design relate directly to the qualities of obsolescence, permanence, and value that we have been discussing. Through manipulation of shroud or sheath design it has been possible to create artificially consumer dissatisfaction on a fashionable or modish level. And this may be just the kind of design which Bill Blau referred to in *Fortune* magazine (February 1968), where he spelled out the doom of the design field. For technology has caught up with the Detroit "stylist" and designer-cosmetician. With increasing rates of invention, accelerated technological change, and greater emphasis on "plug-in-plug-out" componentry, the entire product *an sich* becomes truly obsolete within a very short time, to be superseded by a new generation of technologically improved objects.

If we summarize, we see readily that certain aspects of our Kleenex Culture are unavoidable and, in fact, beneficial. However, the dominance of the market place has so far delayed the emergence of a rational design strategy. Obviously, it is easier to sell objects that are thrown away than objects hat are permanent, and industry has done little or nothing to decide what should be thrown away and what should not. It is also much pleasanter (for shareowners, and vice-presidents in charge of marketing) to sell throw-away things that are priced as if they were to be kept permanently. The two alternatives to the present price system, leasing, or lower prices combined with the customers' investment recovery through meaningful trade-in or "model-swapping," have not been explored. Technological innovation is progressing at an ever-accelerating pace while raw materials disappear. (Thus it is instructive to note that it takes about 850 acres of Canadian timber to print one Sunday's *New York Times*. To which may be added: *The New York Times* Sunday edition sells for 50¢ and contains more paper and typography than an unillustrated novel retailing for $7.95. While the *Times* carries about 500 photographs and drawings in its Sunday edition, and a novel does not, book-binding costs average 22¢ per book. It costs the city of New York nearly 10¢ per copy each week to clean up discarded copies of the Sunday *New York Times*.)

The question of whether design and marketing strategy is possible in these areas under a system of private capitalism remains experimental. But it is obvious that in a world of need, answers to this question must be found.

6

SNAKE OIL

AND THALIDOMIDE:

Mass Leisure, and Phony Fads

in the Abundant Society

Our enemy is self-complacency, which must be eliminated before we can really learn anything.

MAO TSE-TUNG

ALL RIGHT: the designer must be conscious of his social and moral responsibility. For design is the most powerful tool yet given man with which to shape his products, his environments, and, by extension, himself; with it, he must analyze the past as well as the foreseeable future consequences of his acts.

The job is much harder to do when every part of the designer's life has been conditioned by a market-oriented, profit-directed system such as that in the United States. A radical departure from these manipulated values is difficult to achieve.

It is the more fortunate nations, those favored by their geographical position and historical circumstances, that today show a grosser spirit and a weaker hold on moral principles.

Nor would I call these nations happy in spite of all the outward signs of their prosperity.

But if even the rich feel burdened by the lack of an ideal, to those who suffer real deprivation an ideal is a first necessity of life. Where there is plenty of bread and a shortage of ideals, bread is no substitute for an ideal. But where bread is short, ideals are bread. (Yevgeny Yevtushenko, *Precocious Autobiography*.)

All design is education of a sort. It may be education by studying or teaching at a school or university, or it may be education through design. In the latter case the designer attempts to educate his manufacturer-client and the people at the market place. Because in most cases the designer has been relegated (or, more often, relegated himself) to the production of "toys for adults" and a whole potpourri of gleaming, glistening, useless gadgets, the question of responsibility is a difficult one to raise. Young people, teen-agers, and pre-pubescents have been propagandized into buying, collecting, and soon discarding useless, expensive trash. It is only rarely that young people overcome this indoctrination.

One notable rebellion against it, however, did occur in Sweden a few years ago when a 10-day "Teen-agers' Fair" attempting to promote products for a teen-age market was boycotted so thoroughly it nearly got put out of business. According to a report in *Sweden NOW* (Vol. 2, No. 12, 1968), a good number of youths resisted what they considered over-consumption by holding their own "Anti-Fair," where the slogan of the day was "Hell, no, we won't buy!" On the big day, buses collected teens from all over Stockholm and drove them to experimental theaters where special programs of politically engagé films and plays were scheduled and such subjects as world hunger, pollution, and drugs were discussed in workshop sessions. In the kids' opinion, the "Teen-agers' Fair" was just the beginning of a systematic plan to exploit young Europeans by enticing them to want more clothes, cars, and "status junk."

But Sweden (once again) is the exception rather than the rule.

Interestingly, the idea of "pure" design and the moral neutrality of the designer always comes up when designers achieve official status or become salaried or subsidized. It seems like an attempt to affirm the identity of the designer and to protect him against officious interference by managerial groups; unfortunately, it is also self-deception and a hoax perpetrated against the public. To paraphrase freely from Paul Goodman's *Like a Conquered Province,* in America at present the great bulk of the billions of dollars for science and design is for research on extrinsically chosen problems, or even on particular products. Of nearly $20 billion marked by government and corporations for research and development, more than 90 per cent is devoted to last-stage designing of hardware for production. Corporations mark up prices 1,000 per cent in order, they say, to pay for basic research, but much of the research is to by-pass other firms' patents. It is hard to believe that this kind of science or design is disinterested and that promoters are not using the prestige of design and science as a talking point. Goodman concludes: ". . . I have heard distinguished scientists —for instance at the California Institute of Technology— mention their guilt at bilking the government out of money that, they know, is *NOT* going to pay off as the government wishes!"

It might be instructive to see just what would happen if *all* social and moral obligations were to be removed from designer and manufacturer alike. What if the advertising, design, manufacturing, market research, profiteering complex were really to be given free rein? Assisted by their tame "scientists" in psychology, engineering, anthropology, sociology, and intermedia, how would they change, or distort, the face of the world?

Last year, in order to illustrate what might happen if design were to continue completely unchecked, I wrote a brief satirical piece that tried to show how the combination of irresponsible design, male chauvinism, and sexual exploitation might be highly destructive. Entitled "The Lolita Project," it appeared in the April, 1970, issue of *The Futurist*

Advertisement from *Argosy*, February, 1969. A result of irresponsible design.

(pp. 52–55). My satire concerned itself with a proposition that, in a society that views women as objects for sexual gratification, an enterprising manufacturer might well begin tooling up for the production and marketing of artificial women. These plastic women were to be animated, thermally heated, response-programmed units, retailing at around $400, in a vast choice of hair colors, skin shadings, and racial types. It also suggested various improvements upon nature, offered by a Special Products Division that would fill orders for, say: a 19-foot-tall, lizard-skin-covered woman equipped with 12 breasts, 3 heads, and programmed to be aggressive.

To my surprise, I began receiving much correspondence as a result of my article. A Ph.D. teaching social psychology at Harvard has written me four times regarding a license to begin manufacturing. Industrial designers from many countries are still writing, offering me money to go into partnership with them and begin turning out Lolita units. A full-size plastic doll (which the Swedish press and I feel resembles Jackie Kennedy Onassis) is now available in three hair colors for $9.95; the advertisement is reproduced elsewhere in this volume. And at the time of this writing, the December, 1970, issue of *Esquire* magazine is featuring the construction of such women, with a cleverly faked color photograph.

But my plastic-woman article was merely a very slightly exaggerated projection of standard marketing methods and practices.

One early use of project design to support political aspirations is recorded in Jay Doblin's *One Hundred Great Product Designs*. In 1937, aware of its fantastic propaganda value, Adolf Hitler placed high on his list of Nazi priorities the design of a car for everyone. He ordered the creation of a new automobile firm, the Volkswagen (People's Car) development company. In early 1939 the VW plant began in a non-existent town, later to be called Wolfsburg:

> Hitler was convinced that large automobiles—the only type produced in Germany during the early '30's—were designed for the privileged classes and are therefore opposed to National Socialist interests. In the spring of

1933, he met with Ferdinand Porsche to plan such a car for the masses—the *Klein-auto*. Porsche, who had experimented with smaller cars for many years, saw in Hitler's enthusiasm the opportunity to realize a dream. Porsche was one of the most highly regarded automotive engineers in Germany at the time. As chief engineer for a number of automobile companies including Lohner, Austo-Daimler, Daimler-Benz and Steyr, Porsche was ideally suited to the task. He and the Führer agreed that the "people's car" should be a four-passenger vehicle with an air-cooled engine, and average between 35 and 40 miles ger gallon, and a top speed of 70 mph. In addition, Hitler stipulated it should cost the German worker approximately $600 to purchase. A sum of $65,000 was appropriated to underwrite preliminary development costs; Porsche completed the first prototype car about two years later in his Stuttgart workshop.

In the United States, design is not overtly used in a political manner: rather, it unblushingly serves purely profit-oriented clients. But the implicit message of most of this design is one that caters almost exclusively to the wants of the upper middle class.

Design at present operates only as a marketing tool of big business. Industrial design specifically was created during the Depression of the thirties to help industry reduce costs and improve appearance, as previously discussed. And the business community aims towards the strivings of the middle and upper middle class as the most rewarding market. Because of the unusually short time during which industrial design has existed as a discipline, it is a field that in the United States is dominated almost entirely by the men who first created it, or by their immediate successors. In other words, design in the United States is largely run by a middle-aged or elderly design-entrepreneur class who, in C. Wright Mills's phrase, are "independent middle-class persons, verbally living out Protestant ideas in small-town America." If one looks through a recent coffee-table volume entitled *Design in America* confected by the members of the Industrial Designers Society of America, it is appalling to see how dehumanizing and sterile everything looks. In hundreds of pictures the members of IDSA are obviously trying to present their best face to the world; the result is a collection of elitist trivia for the home and anti-human devices for the working environment.

In order to work more intelligently, the whole practice of design has to be turned around. Designers can no longer be the employees of corporations, but rather must work directly for the client group—that is, the people who are in need of a product.

At present the role of the designer as an advocate does not exist. A new secretarial chair, for instance, is designed because a furniture manufacturer feels that there may be a profit in putting a new chair on the market. The design staff is then told that a new chair is needed, and what particular price structure it should fit into.

At this stage, ergonomics (or human factors design) is practiced and the designers consult their libraries of vital measurements in the field. Unfortunately, most secretaries in the United States are female, and most human factors design data, also unfortunately, are based on white males between the ages of eighteen and twenty-five. As the few books in the bibliography that deal with ergonomics show, the data have been gathered almost entirely from draftees inducted into the Army (McCormick), Navy personnel (Tufts University), or Dutch Air Force personnel (Butterworth). Aside from some interesting charts in Henry Dreyfuss's *Designing For People, there simply exist no data concerning really vital measurements and statistics of women, children, the elderly, babies, the deformed, etc.*

At any rate, based on a manufacturer's hunch that a new secretarial chair might sell, substantiated by extrapolating and intrapolating the measurements of Dutch pilots during World War II, and fleshed out by whatever stylistic extravaganzas the designer performs, the prototype chair is now ready. Now begins consumer testing and market research. Stripping this research of all the mystical clap-trap supplied to it by the snake-oil brigade from Madison Avenue, this means that a few chairs are either tested or sold under highly controlled circumstances. They may, for instance, be sold through a leading department store in six "test cities." (These are cities of medium population and median income, and are towns in which money is usually alleged to be "ready for new ideas." San Francisco, Los Angeles, Phoenix, Arizona, Madison, Wisconsin, and Cambridge, Massachusetts, are 5 out of a list of 50 such towns.) Stores in which such testing is carried on are usually the leading shops or department stores in their line, catering to a white middle-class audience. So much for market research.

Consumer testing is frequently done in one of two ways: either secretaries are urged to sit in the chairs for however long it takes them to type one sentence (after which their attention is directed towards the delicious upholstery color and texture), or else the chair is sat upon for hundreds of hours by a *machine* to see if one of the chair legs will break off. Neither test, I would submit, really gets down to the fundamentals: do different secretaries experience major discomfort while working hard, seated in a chair for a series of four-hour periods, stretching over weeks, months, or years? *More crucial still is the fact that almost nothing industry designs and markets is ever retested.* If it sells, swell.

When we work as a cross-disciplinary team to design a better chair for secretaries, who are we designing for? Certainly the manufacturer wants to build secretarial chairs only to sell them and make money. The secretary herself must be part of our team. And when the chair is finished (interior designers, decorators, office planners and architects please note!) there must be *real* testing. Nowadays an "average" secretary is usually asked to sit in the new chair, sometimes even for 5 minutes, and then asked, "Well, what do you think?" When she replies, "Gee, the red upholstery is real different!" we take this for assent and go into mass production. But typing involves 8 hours a day, long stretches of work. And even if we test secretaries intelligently on these chairs, how can we see to it that it is *the secretaries themselves who make the decision as to which chair is bought*? Usually *that* decision is made by the boss, the architect, or (God save us) the interior decorator.

And this turnabout in the role of the designer can be accomplished. Our role is changing to that of a "facilitator" who can bring the needs of the people to the attention of manufacturers, government agencies, and the like. The designer then logically becomes no more (and no less) than a tool in the hands of the people.

As it happens, much has already been said about the inadequacy of chairs for typists. And now, finally, a typist's chair has been designed in which secretaries formed part of the planning team and tested it thoroughly. The chair was designed by a team called *Umweltgestaltung*, of Stuttgart. Ergonomics, which was studied in detail, was handled by Ulrich Burandt and the Institute for Hygiene and Workers' Physiology in Zurich, Switzerland, and the chair is manufactured by Drabert & Sons of Minden, Germany. It is thoroughly documented in *Infordesign*, No. 34 (Brussels). But one can only fear that what American designers like to call "sexed-up" chairs will outsell it when it reaches the market place in this country. And it should be remembered that, in any case, secretaries almost never have any say in their employers' purchase of chairs.

If we switch examples from a secretarial chair to a small refrigerator, then entirely new standards are involved. With present marketing and sales structures, an inexpensive refrigerator is merely a refrigerator that has been stripped of "added features." Should someone suggest that the new

Secretarial Posture Chair, designed by "Team Design," Bohl, Kunze, Scheel, and Grünschloss of Stuttgart. Courtesy of *Infordesign* magazine, Brussels.

refrigerator does not fit the needs of residents of the ghetto, low-income families, and other foreseeable users, market research teams have ready answers. The inner-city blacks we are told, are "too irresponsible to use it wisely," the life style of the low-income families is based on "eroding family structures," etc. In other words, the faults of our society are blamed on its victims. As William Ryan points out in his most recent book, *Blaming the Victim*, this is a splendid cop-out, and a wonderful way to get rid of one's feelings of responsibility.

In American society we are daily prompted to feel that there is inherently something shameful and wrong in being "low income."

Design indeed still has far to go. During 1970/71, an all-European design competition concerned with table settings was held under the sponsorship of *Gruppe 21* in West Germany. The competition bore the title *"Tisch 80-Bord 80."* The most responsible entry from an ecological viewpoint was submitted by a former student of mine, Mrs. Barbro Kulvik-Siltavuori of Finland. Where everything else submitted attempted to pander to style-consciousness, consumerism, and induced aesthetic feelings of "object-love," her submission concerned itself with aspects of recycling.

Her proposal, instead of concerning itself with design in the usual sense, attacked a socially relevant problem of the abundant society. Significantly (for an entry from Finland), it opposed the collection of pretty dishes and handsome glasses and storing of these objects until they either become damaged or are replaced for reasons of manipulated "taste."

It was proposed to restrict dishes (at least for special purposes and population groups) to three parts which satisfy minimal need requirements: a large plate, a small plate, and a mug for liquids. Mrs. Kulvik-Siltavuori suggested salt-glazed red clay as a possible material; another suggestion was plastic. She quite properly felt that differences in glaze, coloration, or size were completely negligible when compared with extreme low cost.

These seasonal dishes come in an aerated plastic con-

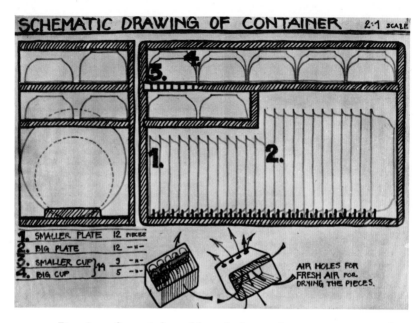

1. SMALLER PLATE   12 PIECES
2. BIG PLATE   12 -"-
3. SMALLER CUP   9 -"-
4. BIG CUP   5 -"-

AIR HOLES FOR FRESH AIR FOR DRYING THE PIECES.

Drawing of aerated washing, drying and storage container for recyclable dishes designed by Mrs. Barbro Kulvik-Siltavuori of Finland. (Courtesy: Barbro Kulvik-Siltavuori and Gruppe 21.)

tainer which is designed to make possible washing, drying and storage of the dishes in the same container. But more importantly, broken dishes and mugs can be returned (like empty beer bottles or recyclable milk bottles) in a garbage sack which is supplied as part of the system. The manufacturer can then use the recycled dishes as raw material: new dishes can be made out of some of the returned plastic, bricks out of the fired clay, etc.

What is important is how the design establishment reacted to her entry. The entry was awarded the *fifteenth* prize (out of fifteen), and the jury remarked: "This concept has considerable originality. . . . At any rate, we know how to appraise the humor of this solution. It is an *amusing* provocation against existing conditions."

As we are taught to equate power, money, and possessions, we deny access to goods to those who are poor or in need. Low income means buying used appliances from Goodwill or the Salvation Army or doing without them entirely. Certainly nothing is designed for the low-income group. The philosophy behind this is that "if only these people had more money," why then they could participate in the "American Dream." In this way we manage to separate out all minority groups—and artificially create quite a few new ones. These minority groups are people who deviate from the arbitrary norms established by the ruling middle-class power structure. By fractionalizing these minorities even further (it seems that there are "educable blacks," "deserving poor," "exceptional children," etc.) and blaming them for the problems which we tell them they have, we have managed to induce in ourselves a specious feeling of superiority and a sense of belonging to whatever Silent Majority is the norm.

It is obvious that the skills of the designer must be made more accessible to all the people. This will mean the restructuring of the role of the designer into that of a community problem-solver. His only allegiance will be to the "direct" clients, the actual users of the devices, tools, products, and environments that he designs. His secondary role will be that of facilitating the production or redesign of these things. When such an idea is proposed to industries, they immediately bemoan the lack of citizen participation. Teachers tell us that parents won't come to PTA meetings; community organizers inform us that citizens won't join their block organization; the tenants, planners tell us, won't come to neighborhood urban renewal meetings. Isn't it interesting to note that *we accuse people of apathy at the very same time they themselves complain about being excluded?*

But probably the most telling point that can be made against the design profession lies in the fact that nearly everything mentioned above is never discussed in either schools or offices. Most designers accept their role as master stylists; they never question the ways in which they help an exploitative system victimize all the people; they fail to realize the divisive role of design in hardening our class structure: they are the "good Germans" of the profession.

Of course, there are certain differences in design as it is practiced and thought about by various age groups and nationalities. Most of the arguments put forward above would be considered just simple common sense by the majority of designers practicing in Sweden, Finland, and most of the socialist countries. In the United States many design students, and quite a few younger designers, feel a tremendous emptiness in their role and an incongruence of what they think to what they do.

Lately some designers in the United States have even been known to permit themselves to *feel* (besides thinking and designing). But most designers still think that their employment as servants of the military-industrial complex is one of the "givens." A curious note of paternalism still dominates design thinking. As the head of one of the largest Chicago design offices said to me at a recent meeting: "We've got to do something for the migrant workers that is good for them—but not too good, or they'll never get off their asses!" When the residents of a low-income area in Lafayette, Indiana, were designing a playground together with architectural students from Notre Dame University a solution was presented in the neighborhood council that better suited the community's desires. "They can't do that, those are *my* niggers!" was the reaction of one of the design students.

Rather than leaving this unresolved series of charges against the design profession, I would like to give an example of how things *could* work.

My daughter, Jenni Satu, is almost six months old now and has learned to sit on the floor, wobble a little bit, and play. This is a good time to acquaint her with books. Since paper or plastic pages have a tendency to cut sensitive little fingers, and since books made for small children and printed on heavy cardboard have pages that are difficult for them to turn, I began looking around for cloth books. There are some. In fact, there are *eight!* and they are all manufactured by the Hampton Publishing Company, of Chicago. Each book consists of a cover, a back, and 3 (count them, *three!*) leaves, thus a total of *6 pages per book.* They sell for

$2 *each!* They are printed in non-toxic colors on cloth. The illustrations, while reminiscent of romantic paintings done about 100 years ago, were actually done in 1935. Under each of the 6 pictures that make up a book there is printed a helpful descriptive phrase such as "BALL." Since most one-year-olds are incapable of reading and, if read to, demand greater verbalization than that, both pictures and text don't apply. A small child is turned on to textures, color contrasts, optical effects, and things it can suck. So one of my

Shown is a commercially available baby book that costs $2. Below it a redesigned book, more suited for a child's needs and estimated to sell for 60 cents. Designed and developed from an idea of the author's by Arlene Klasky, California Institute of the Arts.

students has designed a new book: it has 10 leaves, or 20 pages. One of them is a small pocket with teddy-bear-textured cloth on the inside. Another page is a mirrored cloth surface. Still other pages present simple color spots, optical saturation patterns, textures that feel good and things that go squeak. In addition, the pages are split horizontally so that the child can combine the 10 pages into over 40 patterns. It is still made of cloth. The colors are still non-toxic, and it can now sell for 59¢. But that is not the end of the design process: my student has also set up a jig, so that these books can be made by blind people, either in hospitals or as a sort of "cottage industry" project. Through this design it was possible to combine an advocacy for two groups: to give delight to small children, and also to supply the need for meaningful work for the blind.

As indicated throughout this book, design is discriminatory against major sections of the population. Just by comparing the controls, switches, knobs, and general design of those tools and appliances which in our society seem to be in the province of women ("homemakers") with those that seem to be "male-oriented," we can see vast differences. In spite of the clients' differing age, occupation, sex, schooling, etc., most designers seem to design for an exclusively sexist, male chauvinist audience. The ideal consumer is between eighteen and twenty-five, male, white, middle-income, and if we look at ergonomic data published by designers themselves, *exactly* 6 feet tall, weighing *exactly* 185 pounds. We have seen that the amount of testing done among various population groups is tokenism at best. Furthermore, designers know very little about what people really need or want.

By studying the curricula of 58 schools that teach design, I find that courses in psychology and the social sciences are nearly always absent. When they do exist, they operate under the nomenclature of *Consumer Preferences 201*, *The Psychology of the Market Place*, and *Buying Behavior of Consumer Groups*. There are some such psychology and social science courses taught to latent designers at such places as

Purdue University, University of Southern California, and the School of Design, Illinois Institute of Technology. (It is only fair to point out that the School of Design, California Institute of the Arts, is a notable exception to this: here the mix between social psychology and other behavioral sciences and design is firmly established. But a new danger has arisen: some designers and their students play with pop sociology instead of designing. It is obvious that better solutions to the design problems of the real world will come from young people *skilled in the discipline of design,* not by untrained dilettantes toying with "trendy" radical chic.)

Going beyond the badly designed objects that occupy space in our world, are there any good things that many people can afford? It may be useful to examine what *is* available that is both well designed and reasonably priced. While I was still in school, *Interiors* magazine coined the phrase: "The Chair as a Signature Piece of the Designer." Good or bad, the phrase has stayed. Today the consumer who wishes to buy a chair is faced by a bewildering array of 21,336 different models. Many of these are American, but we also import, from Finland, Sweden, Italy, Japan, and many other countries. There are chairs in production that are careful copies of seating units from predynastic Egypt; other inflatable chairs, incorporating the most recent bits of plastic and electronics, owe their aesthetic debts to the latest moon shot. In between there are faithful reproductions of Hepplewhite, Early American, Duncan Phyfe, and much else, including newly created styles such as "Japanese Colonial," "Plastic Baroque," and the "Navaho Look." Prices vary: it was possible to get an inflatable chair for as little as 59¢; an easy chair now sold that is part Swedish but features Japanese electronics in its stereo head-sets and a German impeller motor for a rippling motion in the back rest, sells for a cool $16,500 each. Aesthetically, as well as for many specific functions of use, or telesic aptness, there are probably at least 500 "good" chairs. But we might concern ourselves with 3 chairs I consider great, 2 of which have stood the test of time so well that most people are astounded when they find out when these chairs first came into being.

The Director Chair in its most current version is a scissor-legged wooden construction with slip-on seats and back, made of No. 8 duck, with a 300-pound test strength. It is extremely comfortable to sit in for long periods of time, and that is quite unusual for a chair without cushions or pads. For storage or ease of shipping, it folds up into a compact package, weighing less than 15 pounds. It has another unusual advantage in that it can serve equally well as an easy chair, desk chair, lounge chair, or dining chair. We use eight of them in our home, their light weight, compactness, and ease of maintenance together with great comfort and low price making them especially attractive chairs for today's greater mobility and changing life styles. At present, the chair can be bought from Sears Roebuck for $12.88. Jay Doblin, in his book *One Hundred Great Product Designs,* calls it ". . . a tremendous buy, probably the best dollar's-worth of furniture available." Most people, when asked to put a date to it, assume that it was designed during the late forties. They are mistaken by one century. The chair can be seen in early French and American photographs and re-appears more frequently in pictures made during the Civil

ABOVE: Director Chair, manufactured by The Telescope Folding Furniture Co., Inc., Granville, New York.

RIGHT: "Lounge Chair" (1938) by Durchan Bonet and Ferrari-Hardoy. Metal rod and leather. Manufactured by Artek-Pascoe, Inc. Collection The Museum of Modern Art, New York, Edgar Kaufmann Fund.

War. In its present form it is produced by a number of firms: the Telescope Folding Furniture Company of Granville, New York, and the Gold Medal Company of Racine, Wisconsin, now make at least 75,000 chairs annually. Estimates made by current producers of the chair put the quantity produced since 1900 in excess of 5 million in the United States alone. Jay Doblin mentions that the Gold Medal Company can trace their present model back to 1903. In addition, there exist British, German, Swedish, and Finnish versions of this chair. The British version, tatted up for present-day consumers in leather and walnut, is sold as the "British Campaign Officer's Chair."

In 1940 Hans Knoll purchased the design of a chair developed by Ferrari-Hardoy and Durchan Bonet. This construction of 2 interlinking, open tetrahedra made of steel rods, covered with a sling of leather or canvas, has since become known as the Hardoy Chair among designers, as the Butterfly Chair, Campaign Chair, Sling Chair, or Egg-Head's Delight, and Safari Chair to the general public. It is an extremely comfortable indoor-outdoor easy chair, completely amphibious when using the canvas sling, lightweight and, while not foldable in most models, at least stackable. The original Knoll-Hardoy retailed for $90 in 1940 with a leather sling. Rip-offs by competing manufacturers brought that price down, at least on the West Coast, to $3.95 in 1950. Overproduction of these rip-offs finally made the chair into a free give-away at some supermarkets in the West and Southwest (with a purchase of $40 worth of groceries). The origins of the Hardoy Chair are obscure, but it was produced in a wooden version that folded, by the ubiquitous Gold Medal Furniture Company of Racine, Wisconsin, as early as 1895. Nearly 7 million Hardoy Chairs, or their copies, have been sold during the last three decades. The reasons for its popularity are identical to those for the Director Chair mentioned above; like it, the chair has resisted any attempt to infuse it with elements of "status" or "elitism."

"The Sack," designed by Piero Gatti and Cesare Paolini, was introduced to the Italian public late in 1968. Essen-

tially, it is a leather-covered sack filled with plastic grains.
The original retail price (in Italy) was $80. The chair is
lightweight, essentially bag-like, and easy to carry. The con-
sistency of the plastic fill is such that it molds itself to the
user's contours. Except for the covering material, the chair
also has no connotations of status. Since its introduction,
copies of it, in various covering materials, have brought the
price down to as little as $9.99. It seems to work better in
fabric, best in the original soft and pliable Italian glove
leather. The internal mix of plastic pellets, with a covering
of vinyl or Naugahyde, is probably the least pleasing because
it.doesn't "breathe." Like the Director's Chair and the Hardoy
Chair, it fits in superbly with today's ideas of casual living.
The disadvantage of The Sack and the Hardoy Chair is that
older people find it hard to get in and out of them. What the
three chairs seem to have in common (in spite of having
been designed over the span of more than a century) is ease
of maintenance, easy storage and portability, no concessions
to status, and a low price. In this connection, it is interesting
to note that, at least in the United States, none of these 3
chairs has sold to low-income markets. The reason is not
obscure: the low-income groups have been successfully
victimized by advertising and TV to feel that these are not
"proper" chairs.

"Sacco" (the Sack) Chair, designed by Piero
Gatti, Cesare Paolini, and Franco Teodoro.

Designers may be far from unanimous in picking these 3 chairs as "good design." The "taste-makers" in our society have a disastrous record in selecting what is good design. The Museum of Modern Art, in New York, is usually credited as being the prime arbiter of good taste in designed objects. To these ends, the museum has caused 3 pamphlets to be published during the last 36 years. In 1934 they published a book entitled *Machine Art*. It is a heavily illustrated guide to an exhibition that was to make machine-made objects palatable to the public, and moreover the museum hand-picked these objects as "aesthetically valid." Of 397 objects thought then to be of lasting value, 396 have failed to survive. Only the chemical flasks and beakers, made by Coors of Colorado, still survive in today's laboratories (after enjoying a brief vogue that was museum-induced, during which the intelligentsia used them as wine decanters, vases, and ashtrays).

In 1939, the museum held a second exhibition; the pamphlet *Organic Design* illustrates the various entries. Of 70 designs, only one, entry A–3501 designed by Saarinen and Eames, saw further development. The box score of the 1939 exhibition was zero. However, entry A–3501 was parlayed into two competing chairs by its two respective designers. The Saarinen Womb Chair of 1948 and the Eames Lounge Chair of 1957 are both spin-offs of that exhibition entry.

In 1950 an international exhibition under the title "Prize Designs for Modern Furniture" was held at the museum. In spite of the fact that most of the entries were made this time by corporate design staffs and furniture companies, only one of 46 designs survives to this day. Since the Saarinen and Eames Chairs mentioned above sell for $400 and $654 respectively, their real impact on people has been negligible. But when we deal with the taste-making *apparat* of the Museum of Modern Art, a score of 3 successes and 510 misses is far from reassuring. What is even more impressive is what the museum has missed: Mies van der Rohe's Barcelona Chair was designed in the twenties. Knoll International revived the chair in the fifties, sold it (only in pairs) at $750 *each:* it has since become the prime status symbol of big business and graces the entrance hall of most gauleiters of industry around the world. Another one missed by the museum, and equally profitably reintroduced by Knoll, was the canvas and steel Le Corbusier Chair, also originating in the twenties.

It is most interesting to compare the many museum catalogues of "well-designed objects." Whether printed in the twenties, thirties, fifties, or seventies, the objects are usually the same: a few chairs, some automobiles, cutlery, lamps, ashtrays, and maybe a photograph of the ever-present DC-3 airplane. Innovation of new objects seems to go more and more towards the development of tawdry junk for the annual Christmas gift market, the invention of toys for adults. When plugging in the first electric toasters in the twenties, few would have foreseen that in another brief 50 years, the same technology that put a man on the moon would give us an electric moustache brush, a battery-pack-powered carving knife for the roast, and electronic, programmed dildos. ("Joy to the World?") But there have been true originators. I can find nothing designed by the late Dr. Peter Schlumbohm that is not supremely well designed, thoughtfully engineered, a complete breakthrough, and unusually attractive aesthetically.

Dr. Schlumbohm was a self-employed inventor who in 1941 designed the Chemex Coffee Maker. The Chemex was

LEFT: "Chemex Coffee Maker" (1941) by Peter Schlumbohm. Pyrex glass, wood, 9″ high, manufactured by Chemex Corp., U.S.A. Collection The Museum of Modern Art, New York. Gift of Lewis and Conger.

RIGHT: "Water Kettle" (1949) by Peter Schlumbohm. Pyrex glass, 11″ high, manufactured by Chemex Corp., U.S.A. Collection Museum of Modern Art, New York. Gift of the manufacturer.

to be prophetic of all Schlumbohm's later designs: a way of doing things better, more simply, and through non-electric, usually non-mechanical, means. By restudying basic physics, he was able to develop a way of making better coffee more simply. Since its introduction in 1941, many copies of the Chemex system have appeared in other countries, notably the "Melitta" in Germany, as well as several Swedish systems. The coffee maker was followed in 1946 by a cocktail shaker, in 1949 by a glass water kettle that boils water faster because of its configuration, in 1951 by an electric "filter-jet" fan, and by many other items such as snow goggles, a dual-purpose tray, etc. Everything designed by Schlumbohm (who died in 1957) was reasonably priced.

It would be both repetitive and boring to list again the many badly designed toys. While we seem to have gotten away from some of the tin soldiers, bombers, and tanks for boys and wardrobed Barbie dolls for girls, we have instead intergalactic robots. Many of them are as destructive as the war toys, and even more inhuman and mechanistic.

One of the more successful Christmas toys during the 1970 holiday season was a little device selling under the name "Dynamite Shack." A small house made of plastic came equipped with a bundle of (fake) dynamite sticks that were connected by a cable to a detonating device. The idea was for the eager children to sneak the bundle of dynamite sticks down the chimney and then press the detonator. After what the manufacturer called a "satisfying bang," the house would seemingly explode into a dozen or more pieces. Of course, these pieces could be reassembled, and the game could be played over and over again. But I must question both the educational and the entertainment value of a toy that teaches youngsters how to blow up buildings.

The number of toys that presently are well designed, inexpensive, and specifically related to the discovery cycles of a growing child is still somewhat small. But a start has been made. Creative Playthings, of Princeton, New Jersey, and Los Angeles, California, markets such toys from all over the world. I should especially like to commend a series of simple wooden toys from Finland.

These were designed to give both pleasure and training in such skills as twisting, turning, threading, pressing, pushing, etc. They are designed by Jorma Vennola and Pekka Korpijaakko. Several summers ago, Jorma Vennola, who was a student of mine, greatly helped in the invention, design, development, and building of the first portable play and training environment for children with cerebral palsy (CP–1). This environment is pictured and described elsewhere in this book. While working on the environment, Jorma Vennola also developed several toys. One of these is his "Fingermajig." As this is probably an ideal design development, let me describe it briefly.

Two plastic halves, each the exact shape and size of one of the halves of an old-fashioned bicycle bell, are connected into a ball-like configuration. Through a series of holes, a number of dowels protrude by about 1½ inches, each. They have backstops so that they cannot slide out. The heart of the device is a small foam rubber ball. When pushed, the dowels will move in and then jump back out again. The toy comes

in 8 bright colors. Children love to play with it. It has a wonderful feel and resiliency. It provides superb exercise of the hand muscles for all children, as well as those with cerebral palsy, some types of paraplegia, and myasthenia gravis. Being extraordinarily simple and non-mechanical, it does not wear out or need repairs. It floats (making it accidentally one of the few well-designed bathtub toys). With its bright colors, it is a wonderful toy for play in the snow.

Best of all: after transportation charges from Finland and duty payments, it retails for 75¢. (Some stores, usually department stores, have lately taken to selling the "Fingermajig" as a toy for $1 in their children's department, and for $5 each upstairs as an "executive pacifier.") While Creative Playthings is to be commended for marketing these toys, much of the credit must go to Kija Aarikka who first began producing and selling them in Helsinki.

There is much that is designed well, incredibly much more that is designed badly, and a truly frightening amount of things that are never designed at all. Now, I am not pleading for more and more products. A world with its back to the wall, ecologically speaking, can ill afford any of the four stages of pollution: the rape of raw materials, the pollution created in manufacturing, the overabundance of products, or the pollution of disused products rotting away. I am frightened when Herman Kahn quotes a song sung

"Fingermajig" (FAR LEFT) and "Threading" (BELOW) toys designed by Jorma Vennola (Finland). "Turning" (TOP LEFT) and "Pushing" (NEAR LEFT) toys designed by Jorma Vennola and Pekka Korpijaakko (Finland). Courtesy of Creative Playthings, Princeton, New Jersey and Los Angeles, California.

every morning by the Matsushita factory workers in Japan:

> Let's put our strength and minds together,
> Doing our best to promote production,
> Sending our goods to the people of the world,
> *Endlessly and continuously.* . . . [my italics]

Nonetheless, there are things that are needed, things that are needed now. Often designers tend not to design something because a better technology seems on its way. But that is a cop-out. When a blind man needs a better writing tool for taking notes in Braille, then it is little use telling him that in 10 years, tape recorders the size of a cigarette pack will cost less than $10. First of all, he needs the writing tool *now;* secondly, present-day monopolistic practices make such projections of future price highly unlikely. After all, it is monopoly agreements and price-fixing that result in a hearing aid consisting of an earpiece and a pocket-sized amplifier which cost $6 to produce, to retail for $470. Because few designers seem to think up and develop the type of products really needed, I have explained about 100 of these in the passage to come.

(I must apologize for listing them in a sort of "designer's shorthand." To describe each and every product would fill many volumes. That others may have had similar ideas is a fact of which I am abundantly aware. However, they have done little or nothing about it. Some of these products have already been designed by some of my students and are illustrated in this book. What we are all looking for are ways of facilitating their production.)

PRODUCTS NEEDED NOW: Health care and disease prevention as well as diagnostics might be a good place to start. Of course, there is a need for intelligently designed heart-lung machines, simplified open-heart surgery, and much else. Instruments nearly that sophisticated, like the drills and saws for osteoplastic craniotomies designed by a student of mine, C. Collins Pippin, are illustrated elsewhere in this book. But much can be done on a much lower, almost "gadgety" level. Take fever thermometers, for instance. None exist at present that give an extremely fast, efficient read-out and are low in

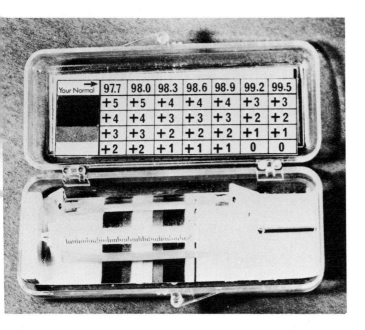

Your Normal →	97.7	98.0	98.3	98.6	98.9	99.2	99.5
	+5	+5	+4	+4	+4	+3	+3
	+4	+4	+3	+3	+3	+2	+2
	+3	+3	+2	+2	+2	+1	+1
	+2	+2	+1	+1	+1	0	0

A color-coded box with a linear magnifier holds the thermometer and permits readings by illiterates. Designed by Sally Niederauer, as a student at Purdue University.

cost. None exist that are color-coded so that people unfamiliar with writing, or with the convention of 98.6° being "normal," can find out what their temperature is. What about a thermometer that makes it possible for a blind person to take and determine his own temperature? One only exists; it is imported from Switzerland, breaks down often, and costs $10.

All of us have wasted too much of nurses' or nurses aides' valuable time as they tried to find our pulse and then take it. Modern electronics could provide us with a pulse-taker that responds to the actual pulse vibrations. It could be about the size of an old-fashioned pocket watch and need cost no more than $15. One of my students, Bob Worrell, has in fact designed such a device. But here we encounter a second level of complexity. Many people who are perfectly capable of using such a device on patients may be functional illiterates and therefore incapable of reading a complicated dial, doing some arithmetic regarding pulse mode, and then transferring this data correctly to a patient's records. Obviously a digital read-out can be devised. To take it a step further, the digital read-out can be linked to a rubber stamp wheel, so that it can be read (but need not be) and can be transferred directly to the appropriate chart. (In fact, labor unions and craft guilds have much on their books that is restrictive in this sense. Gas and electric meters, for instance, present a confusing array of 5 or 7 different dials in many communities. The union and/or the company then forces new men to undergo months of training in how to read these unreadable devices.)

Surely blood pressure can be taken more easily and more comfortably for anxiety-prone patients.

Urinalysis can be a gamble. One commonly used device works like a hydrometer, but because the scale inside the tube is printed on a piece of paper that is not firmly fixed, such readings are completely meaningless. This device sells for under $4. At the other end of the scale are vast electronic machines that do a good job in hospitals. These cost several thousand dollars. Humbert Olivari, a student of mine at North Carolina State, attempted to prove that it was possible to design a small electronic device that could be made for under $30, thus filling the mid-range of urinalysis devices through thermocouples.

Crutches are badly designed; braces are costly and not designed for greater adjustability to differing body proportions. Canes for the blind were recently redesigned for the first time by Robert Senn; they are described and pictured elsewhere in this book.

Exercising vehicles for children with cerebral palsy, paraplegia and quadriplegia, myasthenia gravis, and other debilitating diseases are fully discussed and illustrated in another chapter. Right now I am designing an exercising-and-fun vehicle for children who are only capable of uncontrolled gross body-movements. But what about such devices as exercising tools for middle-aged males facing stroke or coronaries? What about such devices for veterans newly

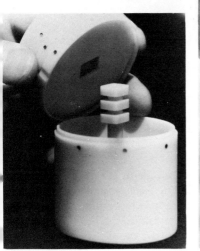 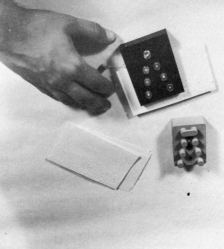

LEFT: Pill box designed to be tamperproof, so that small children are denied access to pills. Designed by David Hausman, as a student at Purdue University.

RIGHT: Writing instrument for the blind that is a major improvement on models now existing. Designed by Solbrit Lanquist and James Hennessey, as graduate students at California Institute of the Arts.

fitted with prosthetic arms or legs? Such vehicles could be stationary, self-propelled, or used in water for hydrotherapy. It is sad to record that when my students and I designed our first vehicles of this sort in the late fifties, these were then the only ones available anywhere. Since then only three badly restructured tricycles have been produced commercially.

There is no diagnostic device for quickly, accurately, and inexpensively obtaining galvanic skin response. (A fountain-pen-sized probe would do the job.)

No really safe pillbox is currently manufactured at a cost low enough so that it can be given away with prescriptions. In the United States more than half a hundred children die annually because they overdose themselves with various pills and capsules. There is no "safe" for medicines kept in the home that is child-proof. There is no "safe" for household cleaners and chemicals that is tamper-proof.

Blind adults need an instrument on which to take notes in Braille. At present they are faced either with using an expensive and bulky typewriter (since they cannot see, the

typewriter needs extra controls), or using a wholly inade-
quate "pocket stylus and slate." This instrument is small
enough to be carried. However, as the impressions are made
in a *downwards* direction and Braille is *raised,* everything
has to be written *backwards.* A team of two graduate students at Cal Arts, James Hennessey and Solbritt Lanquist,
have designed an inexpensive, pocket-sized Braille writer.
But blind people also need more meaningful work than making baskets and brooms. It could become the job of the designer to develop manufacturing processes directly related to
the often impressive skills of the blind.

There are many other groups that we have singled out and
called "deprived," "disadvantaged," or "retarded." By doing
so we manage to blame them for the shortcomings of our
society. Their skills must be investigated to lead to the design and development of things for them to do. It goes without saying that members of the disaffected groups must in
each case be part of the design team.

When the San Francisco Museum of Art recently had a
sculpture show for the blind, many were smilingly impressed by the foresight and "creativity" of the museum's
curator. Hordes of tiny blind children were marched through
the galleries while the flashbulbs busily clicked away. But
surely we cannot deprive the blind, and partially sighted, of
all the heritage of sculpture, using their ailments only for
occasional ego trips by museum directors. (By the way: the
exhibition and the idea for it was borrowed from a show
called "Art for the Blind," held at Moderna Museet in Stockholm in July and August, 1968. That show, however, was
held in a totally blacked-out area, thus temporarily "blind-
ing" even the sighted. All sculpture was identified in Braille;
to go through the show one had to *feel* one's way along a
waist-high rail. When one touched the Braille identification,
a tiny, nearly invisible, wheat-germ bulb would light up,
illuminating only a one-inch area that identified each object
for those of us who could not read Braille. Quite a difference
from the San Francisco show!)

At present we are working on a "Sensory Stimulation
Wall." I am indebted for the idea to a former student of

mine, Charlie Schreiner of Purdue University, and Yrjö Sotamaa of Helsinki. Essentially the wall is a space grid measuring 2 x 5 feet and is one foot deep. "Plugged" into this wall are 10 one-foot cubes. Each "does" things. They squeak, show multi-faceted reflections, are three-dimensional interior "feelies" for the hand to explore, switch on lights, and much else. This unit can live in the nursery or day-care center, lying on one of its 5-foot-long sides. Children as young as one year can explore the unit and play with it. As the child grows older or new skills develop, new types of cubes such as aquariums, rear-projection slide screens, electronic toys, and much else, can be added or substituted. Specific skills such as lacing, buttoning, tying, working a zipper or buckles, snaps, etc., can be taught.

The discomfort, pain, and puzzlement of a small baby that is teething is really pathetic. After experiencing 4½ million years of this (according to Robert Ardrey), we have developed *one* toy: a plastic tube filled with water that can be frozen. It gives the baby comfort for about 5 minutes, by which time it has warmed up and is therefore no longer soothing. Surely we can do better than that.

According to the best estimates there are at present about 150 million people (world-wide) who are bed-ridden, would like to read, but cannot turn the pages of a book. Seven different "page-turners" are available in Sweden, 3 in the United States; none of them work. After designing one, we might also link it to a small overhead opaque projector.

The aged need furniture that is easy to get into or out of. This furniture should be low in cost, easy to clean, and easy to maintain. In the retirement villages of Florida and the West Coast there live hundreds of cabinet-makers, designers, and craftsmen, whose most challenging stimulus is the weekend Canasta tournament. The furniture can be designed and built by the client group involved.

What are meaningful, constructive games for the elderly? Surely shuffleboard is not the only option.

Handicapped people, the elderly, and some children need walking aids. All walkers available at present are dangerous, unwieldy, and expensive. *Any* compassionate and well-trained fourth-year design student could design a better walker than any that are now available, in less than one hour.

It seems that a large number of people whose eyesight is less than perfect cannot wear contact lenses. Eyeglass frames change from year to year. But what about self-adhering glasses that attach to the temples, greater peripheral vision, glasses that change color (for bright sunlight, snow, night driving, etc.) chemically?

An ambulance can cost as much as $28,500. Where are well-designed, low-cost inserts that would convert any station wagon to an ambulance for use during a national emergency? With the number and prices of ambulances now, this particular national emergency began about 20 years ago!

In a later chapter an environmental exercising and play cube that was built for handicapped children in Finland is pictured and discussed. What about other cubes? Experimental cube child-care centers, cubes that can be used under water and mid-water, and knock-down cubes that can be used for play, testing, and diagnostic purposes? When university students (another exploited group) move into an old apartment, they needlessly spend much money to make it inhabitable. The services of such an apartment are often indispensable: running water, a toilet and bathtub, heating, a kitchen complex, windows, and room for storage. Much money and time is spent on painting walls and floors, paint that eventually remains behind as an improvement for the slum landlord. And of course there are many people living in slums who cannot afford any improvements. Interior living cubes can be constructed that will make it possible to combine sleeping, working, and sitting surfaces into an aesthetically manageable entity that uses all of the resources of the apartment itself in an ancillary way only, but shuts them out visually. Friends of mine have constructed three such cubes (one to sleep, eat, and entertain in; one for work; and one as a play environment for the baby; each 8 feet cubed) and installed them in their rambling, ugly, slum apartment in Chicago. Recently they have moved the cubes (knocked-down and packaged flat) to a new, equally expensive and ugly tenement in Buenos Aires.

Nearly two years ago, experiments carried on at *Konstackskolan* in Stockholm and fully documented in the Swedish design magazine, *Form*, have shown that people in wheel-chairs cannot use pay telephones or revolving doors, or buy articles of their choice in a supermarket. Much of this also holds true for people on crutches. But nearly two years have passed now. Where are the pillar-telephones (with an acoustical hush-umbrella) for the handicapped? Where are inexpensive but durable conversion units that will change sidewalk steps and curbs into ramps? Where are the revolving display units for supermarket merchandise?

Thornton Ladd's Pasadena Art Museum, completed in 1970 and accurately described in *Time* (May 24, 1971, p. 68) as "a regrettable hybrid of cruise-ship lounge and California bathroom," makes no concessions to the very young, the elderly, or the disabled. Steps must be used to enter it, to leave it, or, once within, to go to the lower level: there are no ramps anywhere. The colossal doors would defeat anyone on crutches or in a wheel-chair (always assuming they were carried up the steps). Small children in prams or strollers are out of luck, as are their unfortunate parents. The floors are slippery for anyone attempting to maintain a precarious balance with the aid of a cane: elderly people, who make up an unusually sizable portion of Pasadena's population; women who are pregnant and somewhat unsteady; young people with, say, a temporary skiing injury: in short, those who might be expected to have enough disposable leisure time to come to the museum. The ground outside the museum is surrounded by a sort of optical moat: crushed white marble and crystal glitters in the unshaded sun in mind-blinding harshness. After passing this visual booby-trap, even the healthy need quite a while before their eyes adjust enough to be able to see some of the colors of the paintings inside.

In a splendidly engineered, vertical black slum in Chicago, black women are forced to walk a round trip of nearly five miles to shop at the nearest supermarket. Public

transport is unavailable. If a woman is pregnant, she has t
rest her parcels on the head of her unborn baby on the wa
back. The architectonic and hydraulic problems of preg
nancy are also permanent problems of the obese. Simpl
chores such as taking a bath and getting out of bed create
host of imbalances. Yet, the tools of making life easier fo
these people have not been forthcoming.

Highly specialized work often calls for highly specialize
equipment. As a case history: At Cal Arts we discovered tha
dancers and dance students could relax their legs mor
efficiently by elevating them as much as possible. No seatin
unit (with the partial exception of the ill-fated "Barca
Lounger of 1939) exists for this function. By making dancer
and dance students (the client group) part of the desig

A "high-speed" relaxation chair
specifically designed for dancers
By Douglas Schoeffler, as a student a
California Institute of the Arts.

eam, a graduating student, Douglas Schoeffler, developed a relaxing chair that does the job. In the first picture it is shown in a normal seating position. In this mode it can also be used like a rocking chair. In the second illustration it is in a "high-speed relaxation" mode. Just by putting one's arms behind one's head, one tilts the chair to the second position. Many of these chairs have been built and sold to professional dancers or students at cost. Since some of the students cannot afford even $18, they can also purchase plans and templates for 75¢.

Safety and consumer protection could (and does) fill many volumes. What interests me most is how *unsafe* most safety devices are. Safety glasses, goggles, helmets, shoes, and truck drivers' cabs have been mentioned in Chapter Four. In addition, there is a problem with hydraulic dollies, which we use when moving heavy objects in order to escape injury. These are overpriced by, minimally, 600 per cent. Also, noise pollution not only affects hearing, but has been demonstrated to have bad effects on the cardiovascular system. Ear protectors, even when designed for gun enthusiasts, are unsafe, outmoded, and don't work well. There are no efficient smog masks. There are no efficient back supports, groin supports, girdles, and back braces for men in the construction industry. (Excellent designs for these exist in Eastern Europe. However, both our "captains of industry" and their captive designers seem to assume that "nobody ever really works that hard."

One of the most dangerous chunks of equipment available in the United States is the school bus. They are unsafe vehicles, and they give insufficient protection to children and driver. Local school boards seem to believe in cutting corners both literally and figuratively. The excellent German buses that exist for this purpose are not bought, and American transportation firms are unwilling to build a better vehicle. These 30-year-old death traps rattle down the twisting mountain roads of North Carolina, where a local law permits them to be driven by fifteen-year-olds.

There is no well-designed first aid kit on the market.

Most farm accidents occur with tractors. Nearly all of them have no roll-bars.

All farm machinery and farm implements are unsafe.

With the number of boating accidents that affect small children and babies, it is amazing that no "automatic face-up" life vests exist.

There has been much talk about road safety. Bucky Fuller's Octahedral Truss (a structural system) would certainly make a better guard railing than anything now used. A research team of graduate students at Purdue, headed by Michael Crotty, proposed such a system in 1967. It was rejected as too costly. With an average of fifty thousand deaths a year, what does cost mean?

Road services are at fault here, too. In states like Indiana, the beading which connects narrow original roads with later 3-foot-wide increases in width becomes slick and treacherous when iced over.

Apparently, American motorists reject the idea of having a governor attached to their accelerator. But why not attach a device that activates a loud external siren or bell when the speedometer climbs above 70 mph? At least it would give the rest of us fair warning, and a chance to scuttle to safety.

Discrimination against the poor by designers and their employers is reflected in the pricing of many appliances and tools. Rather than designing an object that will sell at a reasonable price and work well, and then adding other choice options to it as the price rises, we seem to delight in a different approach. We have established a new cycle: the cheapest item in the line is usually virtually a toy (Polaroid Colorpack II). One step up in product cost and we reach the level of junk (most mixers and blenders). A further step and we arrive where we should have been at the beginning: an honest piece of equipment, but vastly overpriced (the IBM Selectric Typewriter). But there are still a few more steps to go. The next one is usually the same piece of equipment as before but now "loaded" with extras. This is called *luxury* (any American automobile). Finally, we get basic design simplicity, usually well made and outrageously overpriced. This is status (the Mies van der Rohe Barcelona Chair). In this connection it might be instructive to use a case history.

A number of years ago, Kodak developed a gravity feed system for the slide magazines used in their slide projectors. A projector called the Kodak Carousel was developed. Because the method of handling slides was really excellent, and because the projector itself was of unusually rugged construction, it sold well. But as the dean of the American Industrial Design profession, Raymond Loewy, is so fond of saying, "never leave well enough alone." Soon a new Kodak Carousel model came off the drawing board, the "slim-line" model. Since it was more compact, many people bought it. The basic early model became the Carousel 600 (with push-button slide changing, a choice of lenses, a tray for one size of slide) for $60 (other size slide trays and several lenses available, extra). We proceed next to the 650 (added: accepts several sizes of slide, button changing forward and reverse, remote control forward) for $100; to the 750 (added: remote forward and reverse, a high-low lamp-saver

Two Kodak Carousel Projectors with remote focusing and control chords. The German Kodak Carousel "S" with variable voltage costing approximately $75 and equipped with extra heavy-duty wiring. The American Kodak Carousel Ektagraphic "VA," quite similar but heavier and without the voltage adjustments, awkwardly designed and heavier, $279.50.

switch) for $130; to the 800 (added: remote control focus, built-in timer) for $145; to the 850 (added: automatic focusing rather than remote, a tungsten-halogen lamp, 2 lenses included) for $190; to the 760 (with a larger slide tray than the 850, and a larger lens included) for $149.50; to the 860 (similar to the 850 but with remote focusing) for $200; to the 860QZ (zoom lens included) for $239; as well as a few other intermediary models with different combinations of accessories. The line even includes the RA960 (random access to slides) for $875 and an arc slide projector for $1,500 (with an arc light).

During all this time, however, Kodak has also made point-by-point copies of nearly all of these projectors and sold them under the name "Ektagraphic" to schools and audio-visual departments. The Ektagraphic line cost about $10–$20 more than their consumer's counterparts. There are only two differences between these and the consumer models: Ektagraphic projectors are painted gray rather than black, and they have what Kodak archly calls "sturdier wiring." This means that Ektagraphic projectors (generally not available to the public except through audio-visual stores) have a grounded (3-point plug), heavily insulated wiring, and are less liable to short circuits.

Or in other words: the regular consumer's models, selling from $60 to $239, *are not as safe* as the Ektagraphic line.

Meanwhile, back at Stuttgart, West Germany, Kodak has quietly built and sold their own version, called the Carousel "S." This model *is* safely wired, has its own remote focusing and slide-selection cables and, to top it all, has built-in step-down and step-up transformers that make it usable anywhere in the world, regardless of local voltage. It sells (in Germany) for a cool $75. Kodak of Rochester, New York, actively attempts to discourage Americans from buying it, hinting that it is somehow unsafe or unsuitable. This is, of course, untrue.

The German model communicates simple, safe, and accident-free function both through performance and looks. Should some German consumer wish to try for the miracles of zoom lenses, automatic timing, and what not, these op-

ions are simple plug-in components that can be bought
separately later on. All the accessories that go with the German version of the Carousel, such as slide trays, extra
lenses, etc., are also better designed, more solidly built,
aesthetically more pleasing, and far less expensive. What is
the point here? Well, the Germans seem to be using that
good old American know-how: mass production. They make
just *one* projector with plug-in options, whereas we make
nearly a dozen (counting the Ektagraphic line), all slightly
different from each other and all neatly trapping the consumer with his choice. Our system is designed for consumer dissatisfaction and forced obsolescence. That it is also
expensive and unsafe has been demonstrated.

Of course, projectors need lenses too. In their June, 1971,
issue, the magazine *Modern Photography* conducted a comparison lens test. All lenses were rated "excellent," "very
good," "good," and "acceptable"; they were rated according
to center and edge definition.

The standard Kodak Carousel lens (5-inch F:3.5 Ektanar)
received a rating of "acceptable" (read: "lowest") for center
definition and "good" (read: "second lowest") for the edge
definition. The Kodak zoom lens used in our example above
was tested in three different positions; it was rated "acceptable" four times and "good" twice. By contrast the standard
German Kodak Carousel lens (Projar, F:100mm) has been
tested and rated "excellent" for both edge and center. The
zoom lens for the German Kodak "S" (Vario-Projar, F:3.5,
(70–120mm), has been rated as "excellent" for three and
"very good" for the other three positions.

(It is interesting to note that of the six most popular
lenses available in the United States and tested by "modern
photography," only one could be rated better than "second
lowest": the sole German import on the list.)

It is now more than 40 years since Bucky Fuller designed
a Central Utility Core, a 2-piece molded room that takes care
of most or all of the requirements usually provided by
kitchens and bathrooms. The unit has never really been built
as first envisioned. Craft unions, speculative builders, architects, and designers have all managed to keep it from be-

coming a reality. When Moshe Safdie designed Habitat for Expo '67 in Montreal, he also provided a Central Utility Core. A few hundred of these were built for Habitat, and there it stopped. Certainly a Central Utility Core is needed desperately today.

The Core would revolutionize all housing and it would have even further effects. It is feasible to build a Central Utility Core so that it is possible for people to build the living units themselves out of more organic materials. These would be easier to handle and could be removed when ownership of the house changed. Otherwise, rooms could be provided as hang-on operations to the Central Core.

There is no reason why a combination electric stove and oven has to be larger than a cube 20 x 20 x 20 inches. In fact, we are now developing one which will have as many burners and as much space as "normal" cooking units. Taken in conjunction with our 20-cubic-inch, hand-cranked refrigerator, this gives us a complete cooling, cooking, and baking combination that is less than one fourth the size of the smallest such units now available. It will serve a family of four, easily.

No one has ever designed supermarket shopping carts until recently. The carts that have been designed are made to fit with the concept of impulse buying and not with the needs of the customers. Frequently people steal these carts, which seems a pity since they are so baroquely useless and cost the store nearly $60 each. Since much marketing is done by people without automobiles, the elderly, etc., it would seem that in redesigning a shopping cart one could break the two functions of it into two separate parts. The pushing-rolling action could use better casters, be completely foldable and be used as a free give-away with, say, the first $25 purchase. The storage part can be an equally foldable but recycleable cardboard box. It could recycle from "take-home box" to "garbage pail" or be used as a primary shipping container between supplier and store.

In terms of the Third World, much needs designing. I must repeat that we cannot sit in plush offices in New York or Stockholm and plan things "for them" and "for their own good." Nevertheless, the purpose of this lengthy list is merely to interest people in what can and needs to be done. Power sources, light sources, cooling and refrigeration units, vermin-proof grain storage facilities, simple brick-making and pipe-making systems (for irrigation, waste disposal, etc.), the same kind of inexpensive conversion system that will turn cars and trucks into ambulances that was mentioned earlier: these are some of the needs. But there is much else: communication systems, simple educational devices, water filtration, and immunization and innoculation equipment need design or redesign.

With perfectly useful vehicles such as buses, railroad cars, trains, ferry boats, and steamers lying all over the place and not being used for anything, their redesign as movable class-rooms, vocational re-education centers, emergency hospitals, etc. seems warranted. Old ferry boats might ply some of the tributaries of the Amazon River, for instance, serving as clinics that provide birth control information, abortion counseling, X-rays, prescriptions for eyeglasses, dental care, and treatment for venereal diseases—just to give one possible example.

But most of the needs of the Third World will have to be solved there. Our responsibility as designers lies in seeing that emerging nations don't emulate our own mistakes of misusing design talent as an ego trip for the rich and a profit trip for industry.

At present we are told there is a world-wide food shortage. But the fact remains that there is plenty of food to go around. Food rots, is wasted, or is eaten by vermin. In most of the Third World all perishable food has to be eaten within 24 hours, or before it spoils. There simply are no vermin-proof cooling devices available. Industry (and designers) tend to shrug this problem off and, unconsciously paraphrasing a remark that triggered the first French Revolution, say: "Let them buy refrigerators!" Others become ensnarled in new technologies that *may* someday revolutionize methods of refrigeration.

At Cal Arts one of my graduate students, Jim Hennessey, and I were more concerned as to how Third World people could keep part of their perishables fresh for a week or two starting *now*. We developed a hand-cranked, modular produce cooler. There is a baseboard unit that includes a tire pump, a heat exchanger, a pump, pump valves, and a metering valve, as well as a hand crank. This is surmounted by a 50 cm. styrofoam cube with a lid. It forces hot pressurized air through a heat exchange which returns the air to near-ambient temperature. The air is then metered back into the cooler, where, as it expands, it produces a temperature drop. Other modular cubes can be added. Certainly this is no way to keep two bottles of milk, some Coke, and a roast of beef freezing cold. But 20 minutes of cranking will ensure that, say, a bushel of mangoes will be kept cool enough not to rot (40 F.) for 12 hours. More importantly, the units can be built on village levels in the Third World with existing tools and used valves. Since this problem was solved, we have begun doing research into substituting a sandwich panel (made of two outer layers of used newspapers and a core of dried native leaves) for the styrofoam. The design will be given to UNESCO.

At present no industrial design school is working with agricultural problems (irrigation, pest control, plowing, food storage, etc.). The few design offices that bother with them at all work on "sexier," "zippier" tractors or similar units for the front-lawn market.

There is so much else. American women seem to be willing to explore natural childbirth and Lamaze methods. Yet good graphic information (in the form of slides or simple schematic drawings) doesn't exist. The actual films of childbirth (natural or otherwise) usually induce only a swooning spell in husbands who watch them. Equally needed are simple, graphic, reassuring, non-verbal instructions about abortion.

In the field of transportation we may have to take a large step backward. I was lucky enough to be one of the few people who ever rode as a passenger in the *Graf Zeppelin*. It was both a luxurious and thoroughly delightful experience which colored all my childhood memories of travel. These giant dirigibles consisted of a large passengers' "gondola" which housed the captain's bridge, dining rooms, state rooms, and spacious corridors. The engines were housed in separate "nacelles" that, like the passengers' gondola, hung suspended from the gigantic aluminum structure. They were placed more than 100 feet aft of the passengers' cabins. Both vibration and engine noise were almost nil, and the dirigible, being lighter than air, needed only a slight push to go in the desired direction. Unlike today's jet, it did not rip through the air. In the late thirties, the zeppelins were phased out because of accidents involving the highly flammable helium used. But with our new technology, we may be able to bring them back; we now have gases that are less flammable or inert, thereby eliminating disaster. It would radically reduce pollution across the North Atlantic run, provide a safer and more comfortable trip, and merely add a few hours to the journey. It would be a perfect supplement to today's jets, and certainly a better solution than the criminal delinquency of the proposed SST. The lure of the SST lies in the simple fact that people desperately afraid of flying would like to reduce the duration of their death fear from 8 hours to 3. Dirigibles would give a safer and more comfortable alternative that is ecologically more responsible.

Always the alternative to being able to do things faster is to slow down. To bring back sailing ships for the North Atlantic run is perfectly feasible. The big hang-up with sailing ships was the manpower needed to work the rigging. Today all that could be automated. The second drawback to sailing ships was lack of speed. Today, as we can push people and goods across the ocean in a third of a day flying a jet, an alternate method seems possible here. I am delighted to find that both West Germany and the German Democratic Republic are developing such ships at this time.

While this list is far from complete, there are still a few items of high priority. Virtually nothing is specifically designed for left-handed people. Their problem is more complex than is commonly realized. While (of course) left-handed check books now exist, left-handed booklets of unemployment compensation and welfare forms do not. Some simple tools are bidextrous: a hammer or a screwdriver. To work a Nikon FTN camera, however, (or most cameras) is nearly impossible. Some left-handed people are "right-eyed"; some have a predominating left eye. Steering mechanisms, knobs, and controls in most automobiles are designed for right-handed, right-eyed people.

Some tools designed even for the right-handed majority are also sadly lacking: take typewriters for instance. The keyboard was designed so that the left hand, as well as certain fingers of the right hand, are forced into doing comparatively more work; finger-movement is often awkward. Nonetheless, keyboard organization remains "frozen."

Radical redesign of all sports equipment and especially competition sports equipment is needed. Much is unsafe, and almost everything is so expensive that it is plain why "low-income" people watch baseball on TV instead of going skiing or sailing. Much sports equipment is designed towards status-enhancement. Bad ski bindings on narrow Olympic cross-country skis crack tibias on the beginners' slopes. But beginners' bindings are few, and are merchandised inadequately and apologetically. The National Ski Patrol uses seven different ski rescue sleds. None of them are really safe.

Electric typewriter for use by either left or right hand, and for people with prosthetic hands. It is the size of a standard hardbound book. Designed by Steven Lynch, as a student at Purdue University.

. . .

If my sole intent were to make money, I would develop designs for some of these ideas rather than writing about them. As it is, I am developing designs in those areas that seem to me to have the greatest urgency. The rest are listed to provide turn-ons for others. My feelings about the basic wrongness of patents make this approach consistent and worthwhile.

This may be a good place to mention *Consumer Reports*. It attempts to evaluate products and rejects advertisements in order to operate more independently. The reports are usually intelligent appraisals of user considerations. But because the market is constantly flooded with new items, it is impossible for *Consumer Reports* ever to catch up. While health and safety hazards are pointed out, the magazine writes little about shoddy workmanship and *never* addresses itself to the triviality of an object per se. Thus we may find a 3-page evaluation of competing brands of electric combs, but have to wait 11 months before we find data about the latest frying pans. "Bottom-of-the-line" items, like Woolworth's 29¢ non-electric non-Teflon-coated, smallest frying pan, are almost never written about. Highly specialized pieces of equipment like certain photographic lenses, drafting tools, surveying instruments, medical tools, etc. suffer the same benign neglect. (These same items are usually reviewed in professional magazines. According to a quick survey of some 60 products in 9 professional areas which I made recently, *everything is perfect!*) For how long consumers will submit to risk, harassment and empty promises is an open question, but one which the present depressed market may answer soon. *Consumer Reports*, never addressing itself to the "why" of an object, thus helps us choose the least among several evils.

Packaging can mask all kinds of mistakes, con games, or crimes. When evaluating the Campbell Soup Company's line of frozen breakfasts, Consumers Union, the non-profit group that publishes *Consumer Reports*, said that the food "smelled, felt, and tasted good" and went on to commend the "attractive" packaging. A little later they added that they had found rodent hairs and insect parts in the breakfast sausages. One of the most common packages that assaults us in supermarkets is that of breakfast cereal. Robert B. Choate, Jr., a former consultant on hunger to the Nixon Administration, demonstrated that Wheaties, "Breakfast of Champions," was 29th in nutritional value of 60 dry cereals tested. Nearly half of the cereals tested consisted of "empty calories" and had no nutritional value whatsoever.

Some months ago the British magazine *Design* poked fun at the designers by attributing to them an outlook of: "We are like Gods but we must not let anyone know." Considering all the areas which my list touches upon, it might be easy to assume that I feel that all the problems of the world can be solved through design. But in fact, all I am saying is that a great many problems could use the talents of designers. And this will mean a new role for designers, no longer as tools in the hands of industry but as advocates for users.

PART TWO

HOW IT
COULD BE

7

REBEL WITH A CAUSE:

Creativity vs. Conformity

When you make a thing, a thing
that is new, it is so complicated
making it
that it is bound to be ugly.
But those that make it after you,
they don't have to worry
about making it.
And they can make it pretty, and so
everybody can like it
when the others
make it after you.
PICASSO (as quoted by
Gertrude Stein)

IT IS THE PRIME FUNCTION of the designer to solve problems. My own view is that this means that the designer must also be more sensitive in realizing what problems exist. Frequently a designer will "discover" the existence of a problem that no one had suspected before, will define that problem and then attempt to solve it. This can be read as a definition of the creative process. Without doubt the number of problems that exist as well as their complexity have increased to such an extent that new and better solutions are needed.

At this point I should like to do a number of things: to attempt to describe the need for solving problems, to define that aspect of problem-solving behavior which has been called "creative," and to try to suggest methods for solving problems.

As both a designer and teacher, I am compelled to ask my-

self the question: "How can we make design better?" And
the general consensus seems to be, both in schools and offices in this country and abroad, that the answer does *not*
lie in teaching *more* design. Rather, designers and students
have to familiarize themselves with many other fields and,
by knowing them, redefine the relevance of the designer to
our society. The insights of the social sciences, biology, anthropology, politics, engineering, and technology, the behavioral sciences, and much else, must be brought to bear
on the design process. Ways of doing this are suggested in
great detail throughout this book. But the most important
ability that a designer can bring to his work is the ability to
recognize, isolate, define, and solve problems.

A little over a decade ago, the word "creativity" became a
fashionable cliché for this activity. In fact, one California
university is offering a course entitled "Remedial Creativity
201"!

How and why did being "creative" become a cliché? Certainly the ability to solve problems has been an inherent and
desirable trait throughout human history. Mass production,
mass advertising, the operation of the media, and automation are four contemporary trends that have emphasized
conformity and in this way made creativity a harder ideal to
attain. In the twenties, Henry Ford, attempting to reduce
the price of his cars through standardized production
methods, is reputed to have said, "They [the consumers] can
have any color they want as long as it's black." This implies
that, through curtailing color differences, the price of individual automobiles will be lowered by some $95; con-
versely, consumers must be persuaded that black is a desirable color to have.

The spirit of conformity has accelerated at an amazing
rate. The demands on the individual to conform come from
all directions: not only do the national, state, and local
governments enforce certain standards of behavior, but
there are pressures from neighbors in suburban areas, conformist trends in school, at work, in church, and at play.
What happens if we are unable to operate in so aggressively
conformist an environment? We "blow our top" and are

taken to the nearest psychiatrist for help. The first thing this specialist in human thought and motivation may want to say to us (if not in so many words), is "Well, now, we must *adjust* you." And what is adjustment, if not another word for conformity? This is not to argue for a totally non-conformist world. In fact, conformity is a valuable human trait in that it helps to keep the entire social fabric together. But we have made our severest mistake in confusing *conformity in action* with *conformity in thought.*

Extensive psychological testing has shown that the mysterious quality called "creative imagination" seems to exist in all people but is severely diminished by the time an individual reaches the age of *six*. The environment of school ("You mustn't do this!" "You mustn't do that!" "You call that a drawing of your mother? Why, your mother only has *two* legs." "Nice girls don't do things like that!") sets up a whole screen of blocks in the mind of the child that later inhibits his ability to ideate freely. Of course, some of these prohibitions have social value: moralists tell us that they help the child establish a conscience; psychologists prefer to call this the formation of the superego; religious leaders call it "a sense of right and wrong," or "Soul."

However, society can go amazingly far in attempting to create greater conformity and protect itself from what the current mainstream of culture is pleased to call "deviants." In 1970 Dr. Arnold Hutschnecker suggested in a memo to President Nixon that all children between the ages of six and eight be tested psychologically to determine if they *might* have the kind of tendencies that would turn them towards becoming criminals later in life. The underlying suggestion was that some of these children be tranquilized heavily and maintained in that condition, much as millions of elderly patients in retirement homes are kept under permanent doses of heavy tranquilizers in order to make the work of the nursing staff easier. Unfortunately, this proposal is characteristic of the kinds of pressures towards conformity that are often found in our institutions today.

Too many blocks can effectively stop problem-solving, and the wrong kind of problem statement can do the same. The old saying, "Build a better mousetrap and the world will beat a path to your door," is a case in point. What is the real problem here, to *catch* mice, or to *get rid of them*? Supposing my city is overrun by rodents. Suppose I *do* invent a better mousetrap. The result: next morning I will have 10 million captured mice and rats to contend with. My solution may have been highly innovative, but the original problem statement was wrong. The real problem was to *get rid* of the mice and rats. As a fantasized solution, it might have been better to broadcast an ultrasonic or subsonic beam over every radio and TV set for a few hours, which, while harmless to other living creatures, would sterilize all rats and mice. A few months later the rodent population would be gone. (This raises the ethical question as to whether rats and mice should be permitted to watch television.) More seriously, it would raise the environmental question to what extent some small rodents are important links in the ecosystem.

However, most problems requiring immediate and radical new solutions lie in areas that are quite new.

Chad Oliver, in his science fiction novel *Shadows in the Sun*, says,

> . . . he had to figure it out for himself. That sounds easy enough, being one of the familiar figures of speech of the English language, but Paul Ellery knew that it was not so simple. Most people live and die without ever having to solve a totally new problem. Do you wonder how to make the bicycle stay up? Daddy will show you. Do you wonder how to put the plumbing in your new house? The plumber will show you. Would it be all right to pay a call on Mrs. Layne, after that scandal about the visiting football player? Well, call up the girls and talk it over. Should you serve grasshoppers at your next barbecue? Why, nobody does that. Shall you come home from the office, change into a light toga, and make a small sacrifice in the backyard? What would the neighbors think?
>
> But—how do you deal with a Whumpf in the butter? What do you do about Grlzeads on the stairs? How much should you pay for a new Lttangnuf-fel? Is it okay to abnakave with a prwaatz?

Why, how silly! I never heard of such things. I have enough problems of my own without bothering my head with such goings on.

A Whumpf in the butter! I declare.

A situation completely outside human experience . . .

We live in a society that penalizes highly creative individuals for their non-conformist autonomy. This makes the teaching of problem-solving in design both discouraging and difficult. A twenty-two-year-old student arrives at school with massive blocks against new ways of thinking, engendered by some sixteen years of mis-education, a heritage of childhood and pubescence of being "molded," "adjusted," "shaped." Naturally, he will shop around for a school and study program which seems to hold out the greatest immediate personal rewards. Meanwhile our society continuously evolves new social patterns that promise a slight departure from the mainstream, but without ever endangering the crazy-quilt pattern of marginal groups that make up society as a whole.

First of all, we must understand the psychological aspects of problem-solving. While no psychologist or psychiatrist can as yet point to the exact mechanics of the creative process, more and more insights are becoming available. We know that the ability to generate new ideas freely is a function of the unconscious, and that it is the associative faculty of the brain that is at work here. The ability to come up with many new ideas is inherent in all of us, regardless of age (with the exception of senility and anility) or so-called IQ level (always excepting true morons).

In order to be able to associate freely, multi-disciplinary ability helps. The quantity of knowledge, the quality of memory and recall can enrich this process. The ability to look at things in new ways is indispensable. This "new way of looking at things" can be enhanced through the knowledge and thorough understanding of a second language. For the structure of languages gives us ways of dealing with and experiencing realities, each discretely different in each language.

It is perfectly possible for instance to say, "I am going to

San Francisco," in English. Verbally the same statement
("*Ich gehe nach San Francisco*") is possible to frame in
German, but it makes no sense whatsoever, linguistically. A
qualifier must be added in German, for instance: I am
flying to San Francisco, I am *driving* to San Francisco, etc.
In Navaho and the Eskimo languages the statement must
be even more specifically qualified to make sense: "I (alone,
or with two friends, or whatever) am driving (sometimes I
will drive, sometimes my friend will drive) (by cart, by sled)
to San Francisco (then I will return and my friend will
drive on)." By bringing more than one language to bear on
a problem, we obtain more insight. Whether the language
studied is German, Finnish, Swahili, Piano, Violin, Fortran,
or Cobol matters little.

Intolerance creates more powerful blocks. Within the
social scene, "tolerance" is imperative to the ability to solve
problems. The folk-mind has anticipated the research findings of psychologists: "His mind is in a groove," or "He is in
a real rut," are precise definitions of what really happens. If
someone says "Black," "Jew," "Commie," "Hippie," "Catholic," or "hard hat," or what have you, and the immediate
reaction is "son of a bitch," this has happened. The associational response of the brain has literally worn a groove (or
rut) into the engram-response pattern of the cerebral cortex. Just as Pavlovian psychologists seem impelled by now to
ring a bell every time a dog salivates, an intolerant person
operates on a conditioned-reflex level.

Someone who routinely solves problems also responds to
the concept "security" in a different way from that of his
conformist contemporaries. In 1958 research conducted at
U.C.L.A. among artists, architects, engineers with an unusually large number of patents, composers, musicians,
writers, research scientists doing breakthrough work, has
shown that one characteristic all of these people seem to
have in common (regardless of their financial status) is
that almost all of them are *under*-insured compared to standards set by the population at large. Creative individuals
usually attempt to find security within themselves, rather
than by paying $18.95 a month.

Until they enter school, most people seem to be about equally adept at solving problems. Then the inherent ability to create becomes inhibited by perceptual, cultural, associational, and emotional blocks.

Perceptual blocks are listed here only in order to point out their existence. A dichotomously color-blind person, for instance, has a slight perceptual block in the area of seeing. Trichotomous color-blindness constitutes a more serious block, whereas glaucoma, cataracts, and other phenomena leading to total blindness constitute total perceptual blocks to seeing. Deafness is a complete perceptual block to hearing. Often the psychological inability to use all the senses in observing data will lead to complete blocking. These perceptual blocks, if curable at all, are entirely in the province of the doctor, the surgeon, the psychiatrist.

Cultural blocks, as the name implies, are imposed upon an individual by his cultural surrounding. And in each society a number of taboos endanger independent thinking. The famous Eskimo nine-dot problem which can befuddle the average Westerner for hours on end is solved by Eskimo children within minutes because Eskimo space concepts are quite different from ours. Professor Edward Carpenter explains how the men of the Aklavik tribe in Alaska will draw reliable maps of small islands by waiting for night to close in, and then drawing the map by listening to the waves lapping at the island in the dark. In other words, the island's shape is discerned by a sort of primitive radar. In Eskimo

THE ESKIMO NINE DOT PROBLEM

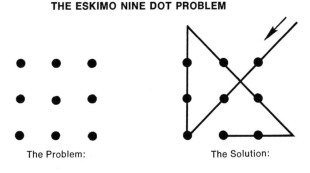

The Problem: The Solution:

art we are sometimes confused, for we have lost the Eskimo's primitive ability to look at a drawing from all sides simultaneously.

Eskimo print: "Spirit [*Tornags*] Devouring Foxes." Author's collection.

Another problem of cultural blocking has been stated by a manufacturer of toilet bowls as follows: while the average American changes his automobile every 2½ years, gets a new suit about every 9 months, buys a refrigerator every 10 years, and even changes his residence about every 5 years, he never buys a new toilet bowl. If one could design the sort of bowl that would make people want to "trade in" their old one, this industry would benefit greatly. At first sight this seems to be a phony job, calling for artificially created obsolescence. And two answers to it immediately jump to the mind of the "stylist." The "Detroit approach": possibly providing the bowl with tailfins and vast chrome ornamentation. Another would be the "Toilet bowls are fun" approach: imprinting the surface with, say, little flowers, birds, or what have you. But intelligent research soon showed that *all toilet bowls are too high* (medically speaking). Ideally, people should assume a lower, squatting position when using this utensil. This can be achieved in two ways: by raising the floor or lowering the bowl. As the client was in the business of manufacturing toilet fixtures, a new lower bowl was designed and built for him. In spite of the obvious medical and sanitary advantages of this, in spite of the fact that now there existed a real reason for buying new toilet bowls, the design was rejected. The manufacturer felt that in this area the cultural block in the public mind was too great, and that it would be impossible for him to advertise his product. This is not an anecdote in somewhat poor taste, but rather an example of a very definite cultural block: the product could have been advertised easily in, say, Finland or Japan.

Cultural blocks operating in this area can be extremely counter-productive. On Earth Day, 1970, it was suggested that people might place 2 or 3 bricks in the water tanks of their toilets and thus cut down on the amount of water used each time. Here again I was able to suggest a redesign. Because what one does while sitting on a toilet differs in both quantity and quality, it seemed simple to redesign the apparatus so that one could select whether a great deal or only a minimal amount of water was needed for flushing. This concept again was rejected by my client—a man who makes his living manufacturing toilet bowls—as being "in bad taste."

Here again the temptation for the designer is just to go ahead and design a product such as the one outlined above, and in this way encourage consumerism. A better strategy is to give the public a series of comparable choices. In the above case this meant providing the redesigned water-saving toilet bowl for those consumers (such as construction firms, housing developers, etc.) about to buy one. At the same time an insert was designed that would have been marketed for under $10 to modify existing bowls to use less water. Finally, the option of just sticking 2 or 3 bricks in the water tank would still have been open to people so inclined.

The cultural taboo about elimination processes has made other developments difficult as well. Toilet tissue is made of paper so constituted that an enormously large quantity of water has to be used in its manufacture. For reasons now obscure to anyone, rolls of toilet paper are of a given width. By reducing this by one inch, millions of gallons of water would be saved daily in the manufacturing process, without cutting down on the function of the tissue itself. Yet here is another idea that is ecologically sound but has gone begging.

If I now try to escalate these examples from the disposal of fecal matter to its constructive use, still more people are turned off. Whenever the concept of recycling body wastes is brought up (for instance in a discussion of space capsules or space stations), people become disturbed. It is useful to remember that, on Liferaft Earth, everything we breathe, drink, eat, wear, or use has gone through billions of digestive systems since the planet was first formed. Our cultural blocks in this area tend to affect our thinking, our thinking affects our acts. We think of streams and lakes as "polluted by urban wastes," we use words like "sludge" and are appalled to find that our water sources are "poisoned" by human excrement. We are confused, as with the better mousetrap mentioned earlier, about whether we want to get rid of excrement or just separate it from our drinking supply.

I am suggesting that the entire field of *anaerobic and aerobic digestion* has been completely neglected. At the time of this writing (December 1970), only three major scientists are involved in studying the entire methane-generating process. Aside from occasional paragraphs in *The Whole Earth Catalog* about solitary British eccentrics who manage to power their automobiles from chicken droppings, the public is largely unaware of the gigantic energy sources that can be mined from our bodily processes of putrefaction, digestion, and waste-making. Yet the recycling of this energy, it seems to me, would be the first logical step in establishing a new life style.

It is well within the ability of contemporary research technology to develop a prime energy converter which, by using anaerobic digestion systems, would make a house truly independent of all external connections. In looking through the newspapers of communes and alternate societies, I have always thought it pathetic that much of their gear (transformers, pumps, high-fidelity components, light generators, projectors, etc.) still has to plug in somewhere. The use of biological recycling for energy would not only make true independence possible, but also bring about a breakthrough in ecology.

It is curious that virtually no research is going on in this area. It is unimportant whether the research is lacking because the field of study is simply too vast, or whether there is some gigantic conspiracy of power among oil companies to suppress such study. The point is rather that we are dealing with an area that the public has been taught to think of as "filthy," and hence, inquiry is aborted by a cultural block.

Much of this has already been tried, but usually just on an individual level. Dr. George W. Groth, Jr., maintains 1,000 hogs in confinement on his farm near San Diego, California. The hog manure operates a 10-kilowatt war-surplus generator, which provides all the electricity needed for both light and power. The liquid manure pit has been capped, and the sewer gas is tied to a gas engine. Hot water from the engine's cooling system runs through 300 feet of copper tubing coiled inside the pit. A temperature of between 90° and 100° is maintained, which provides the best temperature for maximum "digestion." A tiny pump, running off the fan belt pulley, circulates the water. A complete digestion cycle takes about 20 days, but once the process is an on-going one, it is also continuous. Besides providing electric power, the system has virtually no odor, and attracts no flies. Finally, the manure at first breaks down into simple organic compounds like acids and alcohols. Ultimately, as there is no air, it breaks down into water, carbon dioxide, and methane gas.

Experiments of this sort have also been tried in Asia and Africa.

It seems clear that this design strategy can give us a way of using up human and animal waste by converting this material into power sources and recycling what is left. It is curious that what little has been written about it so far has appeared mostly in the underground press and alternative life-style papers.

Associational blocks operate in those areas where psychologically predetermined sets and inhibitions, often going back to our earliest childhood, keep us from thinking freely. An experiment, well known and several years old, will illustrate this point.

In one of our Eastern colleges a six-foot-long steel pipe with a diameter of 1½ inches was immovably fixed into the

cement floor of a basement, so that one foot of the pipe was
below floor level and 5 feet stuck straight up. A ping pong
ball was then introduced into the pipe, so that it would rest
at the bottom, 6 feet from the top opening. Placed in the
room were a miscellaneous collection of tools, utensils,
and gadgets. One thousand students were introduced into
the room, one at a time, and asked to find some method for
getting the ping pong ball out of the pipe. The attempts to
solve the problem were as various as the students themselves: some tried to saw through the pipe, which proved
too strong; others dripped steel filings onto the ping pong
ball and then went "fishing" for it with a magnet, finding
that the magnet would adhere to the pipe wall long before
it could be lowered all the way down. The attempt was
made to raise it with a piece of chewing gum on a piece
of string, but enough pendulum action was acquired in
raising it so that the ball would inevitably drop off. To
stick a series of soda straws together and try to "suck" it up
also proved impossible. But sooner or later almost all of the
students, 917 out of 1,000 (a respectable performance indeed) found a mop and a bucket of water in a corner,
poured the water into the pipe, and floated the ball to the
top. This, however, was only the control group.

A second series of 1,000 students were then asked to solve
the problem again; conditions remaining unchanged with
one slight exception. The bucket of water was removed, and
the psychologists substituted an antique rosewood table on
which a finely cut crystal pitcher of water, two glasses, and
a silver tray rested. Out of the second group only 188 solved
the problem correctly. Why? Obviously because over 80 per
cent in this group failed to "see" the water. The fact that a
crystal pitcher standing on a rosewood table is more notice-
able than a pail in a corner is obvious. What we mean to
imply is that the second group failed to make the associational link between water and a flotation method. The associational gap was a much more difficult one to make with
the handsome pitcher than with the pail, even though
normally we are not given to pouring water out of a bucket
to float ping pong balls either.

Shortly after the end of World War II, Raymond Loewy Associates designed a small home fan and succeeded in making the action truly noiseless. To their consternation, consumer response soon forced this design organization to introduce a new gear into the fan that would give off a slight sound, since the average American housewife associated noise with cooling action and felt that a totally noiseless fan did not provide enough cool air.

Sometimes, the specific training that people have gone through professionally may establish even stronger associational blocks. When faced with the front elevation and the right side elevation of this object, and asked to draw a correct plan view or perspective, architects, engineers, and draftsmen usually fail at a higher rate than that among people untutored in these fields. The correct answer to this particular problem is interesting for another reason as well. Two answers are equally correct, and, depending on which one is given, it becomes possible to see whether the student

OBJECT VISUALIZATION PROBLEM

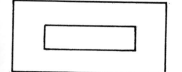
Front Elevation

Right Side Elevation

has arrived at a solution through a species of creative analysis or "sudden insight." The reasoning behind giving answer number one runs somewhat as follows: "The right side elevation does not really work; it should be a center section. Therefore, since it would work perfectly as a center section, I must find a figure where the theoretical center section and the right side elevation are identical. After selecting an equilateral triangle as the answer, I see that the front edge will show up as a line in the front elevation. By rounding this off, the line disappears and the problem is answered

OBJECT VISUALIZATION SOLUTIONS

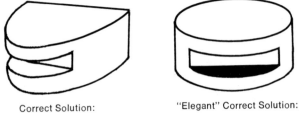

Correct Solution:
Deductive Reasoning

"Elegant" Correct Solution:
Sudden Insight

correctly." In the second case the equally right (but mathematically speaking much more "elegant") answer is arrived at through sudden insight and intuition.

Needless to say, the particular associational block that keeps people from answering this problem correctly, using either solution, lies in the fact that they set up a totally false, not specified, 90° angle relationship and therefore visualize the figure as being rectangular or square. "Rectangularity" or "squareness," then, is the basic block which the solver himself has built into the problem.

Emotional blocks may be the most difficult to overcome. The fear of making a mistake, the fear of making a fool of oneself, a pathological desire for security with an attendant unwillingness to gamble or pioneer, lack of drive to carry an idea through, because of the possibility of eventual failure—all these fall into this category. Other blocks in this area are a deep-seated feeling of inferiority in regard to creation—forcing the designer to "grab the first idea" instead of exploring several alternate solutions—fear of fellow designers, teachers, students, etc.

These points will recapitulate what has been established so far:

1. With constant pressure towards less individualism and greater conformity forced upon our society by mass advertising, mass media, mass production, and automa-

tion, the ability to solve problems in new and unexpected ways is becoming increasingly rare.

2. In a fast-accelerating, increasingly complex society, the designer is faced with more and more problems that can be solved only through new basic insights.

3. Design graduates leave our schools with some know-how, a great many skills, and a certain amount of aesthetic sensitivities, but with almost no method for obtaining any basic insights.

4. They find themselves unfit to solve new problems because of perceptual, cultural, associational, and emotional blocks. These blocks are the direct result of the constantly accelerating rat-race towards conformity and so-called "adjustment."

5. This race is not only inimical to all true design creativity but, in a wider sense, violates the very survival characteristics of the human species.

6. The various blocks are not inherited parts of the personality structure but rather learned, limiting, and inhibiting factors.

Our job then becomes one of establishing methods of doing away with these blocks. By repeatedly facing students and young designers with problems far enough removed from everyday reality so as to *force* them into entirely new thinking patterns, new cortical associations (both feet firmly planted on a pink cloud), by constantly pointing out to them the nature of the various blocks, it is possible to help them realize their creative design potential. By forcing them into solving problems that have never been solved before, lying outside of normal human experience, a habit pattern is slowly established, a habit pattern of solving problems without the interference of blocks (since, with problems removed from everyday experience, blocks cannot operate) and these habit patterns are then carried over into the solving of *all* problems, familiar or not.

What constitutes a "totally new problem, outside of all previous human experience"? If we are asked to design some fabulous animal unlike any we are familiar with, we will probably end up with something possessing the body of a

horse, the legs of an elephant, the tail of a lion, the neck of
a giraffe, the head of a stag, the wings of a bat, and the sting
of a honeybee. In other words, we have really put a lot of
familiar things together in a totally unworkable, unfunc-
tional, unfamiliar way. This is *not* solving a problem. If on
the other hand, we are asked to design a bicycle for a man
with three legs and no arms we can now solve a specific
functional problem far enough removed from everyday previous experience to become valuable in this context.

The late Professor John Arnold, first at M.I.T. and later
at Stanford, pioneered in this field with students in engineering and product design. Most famous of his problems is
probably the "Arcturus IV" project: here the class is given
voluminous reports regarding the inhabitants of the fourth
planet in the Arcturus system, as well as the planet itself.
An extraordinarily tall, slow-moving species descended from
birds, these mythical inhabitants possess many interesting
physiological characteristics. They are hatched from eggs,
possessed of a beak, have bird-like, hollow bones, with three
fingers on each hand and three eyes, the center one of which
is an X-ray eye. Their reaction speed is almost ten times as
slow as that of human beings, the atmosphere they breathe
is pure methane. If a class is now asked to design, say, an
automobile-like vehicle for these people, important and
totally new limits within which to design are immediately
established.

Obviously a gasoline gauge is unnecessary, since the
Arcturians can always see through the gas tank with their
X-ray eye. What about a speedometer? Obviously top speed
will have to be in the neighborhood of 8 miles per hour since
otherwise, with their slow reaction speed, the danger of
crashing into another vehicle before being able to react is
always present. Perceptually, however, such a people would
experience the gradations of speed (up to 8 miles per hour)
much as we experience the speed range in our own automobiles. The answer here then seems easy: subdivide a
speedometer dial. But what kind of a numerical system
would people use who have three fingers on each hand and
three eyes: decimal, duodecimal, binary, sexagecimal? As

these vehicles will be built on earth and exported to Arcturus
IV, should they use a standard gasoline engine shielded
against a methane atmosphere, or must a new type of
engine, specifically designed to operate optimally in methane,
be designed? What of the overall shape of the vehicle?
Should it be egg-shaped (a simple and sturdy form resolved
when aerodynamics are of no importance), or would putting
the Arcturians into an egg-shape become, psychologically
speaking, a return to the womb, lulling them into a false
feeling of security, and, therefore, be imposing the worst
possible shape in terms of vehicular safety? Maybe our design consideration then becomes one of a shape as unlike an
egg as possible—a difficult order to fill indeed!

Arcturus IV is just one of many problems evolved by Professor Arnold and, from the all-too-brief analysis of some of
the possible approaches to it, it will be seen that, while fantastic and science-fictional in content, it is nonetheless a
serious approach to creative problem-solving.

An even richer and more fertile area for problem statements can be derived from nature. In Chapter Five, I discuss
the use of artificial maple seeds as part of a soil-erosion control device. In Chapter Nine, the flight and spiraling behavior of various seeds will be discussed in greater detail.
For now it will suffice to state that a maple seed released in
the air will at first plummet straight down for a foot or so,
and then begin to drift gently and slowly down in a spiraling
motion, finally coming easily to rest on the ground. (During
the spiraling-down stage it is also affected by lateral wind
drift.) Each student was given a maple seed and told to
study it for two weeks. At the end of that period he was told
to find a practical design application for it, utilizing both its
general form (but not size) and its dynamics of downward
motion.

Answers included a 54-inch sustained trouble flare for air-
sea-rescue operations at night, devices to be dropped over
northern Ontario for fish-stocking inaccessible lakes, reforestation capsules, devices for getting food to snow-isolated
cattle and wildlife. Other solutions included toys, a rotating
lab for experiments in induced motion sickness and space

medicine, high-altitude photo reconnaissance rockets in which the cameras form the pod component of a maple seed and a plastic, chemical-filled maple seed, thousands of which would drift into and extinguish inaccessible spots in forest fires.

It can be seen from the foregoing that the "how" in teaching design creativity must consist largely of establishing a milieu in which new approaches can flourish. What has been the function of the school and education in general in this context? It has presented the cultural status quo of its time by dispersing whatever mass of data is currently acceptable as "truth." It has never concerned itself with the *individual* human brain; rather, the tremendous variation in human minds has been taken into account only as something to be flattened out so that the particular curriculum or theory in vogue at the moment can be "sold" with minimal effort. We have failed to recognize that discovery, invention, original thought are culture-smashing activities (remember $E = mc^2$?) whereas so-called education is a culture-preserving mechanism. By its very nature, education, as of now, cannot sponsor any vital new departures in any facet of our culture. It can only *appear to do so* to preserve the sustaining illusion of progress.

One of the major problems of successfully utilizing creative imagination is that "newness" often implies experiment, and experiment implies failure. In our success-oriented culture, the possibility of failure, although an unavoidable concomitant of experiment, works against the matrix. The creative designer, then, must be given not only the chance to experiment, but also the chance to fail. The history of all our progress is littered with a history of experimental failures. This "right to fail," however, does not absolve the designer from responsibility. Here, possibly, is the crux of the matter: to instill in the designer a willingness for experimentation, coupled with a sense of responsibility for his failures. Unfortunately, both a sense of responsibility and an atmosphere permissive to failure are rare indeed.

A more ideal creative-design environment will consist of habituating designers and students to work in areas where heir many blocks and inhibitions cannot operate, and this will imply a high tolerance level for experimental failure. Furthermore, it must mean the teaching and exploring of basic principles which, by their very nature, have no immediate application. This calls for a "suspension of belief" in ready answers, and in the glib, slicked-up *Kitsch* that characterizes most of the design work coming out of schools and offices.

Unfortunately, our society is so structured that all of this is easily possible, paradoxical as this may sound. We need not journey to Arcturus IV to face designers and students with something completely outside their familiar experiences. All we have to do is to design for low-income families. For while designers have addressed themselves to the fads of the middle and upper bourgeoisie, it has also lately become fashionable to do a little bit of token designing for selected "house niggers" representing the poor. Meanwhile, we have lost sight of the fact that a very substantial part of our population is discriminated against in a more subtle fashion.

I am questioning, then, the entire currently popular direction of design. To "sex-up" objects (designers' jargon for making things more attractive to mythical consumers) makes no sense in a world in which basic need for design is very real. In an age that seems to be mastering aspects of form, a return to content is long overdue.

Much of what is suggested throughout this volume in the way of alternate areas for attack by designers also has the useful quality that it will be new to designers and students alike. If (within the meaning of this book) we do that which seems right, we will also develop our ability to see things in a new way and to do things that are new.

8

HOW TO SUCCEED IN
DESIGN WITHOUT
REALLY TRYING:
Areas of Attack for
Responsible Design

> *One cannot build life from refrigerators, politics, credit statements, and crossword puzzles. That is impossible. Nor can one exist for any length of time without poetry, without color, without love.*
>
> ANTOINE DE SAINT-EXUPÉRY

INDUSTRIAL DESIGN differs from its sister arts of architecture and engineering. Where architects and engineers are hired to solve problems, industrial designers are often hired to create new ones. Once they have succeeded in building a new dissatisfaction into people's lives, they are then prepared to find a temporary solution for it.

The basic performance requirements in engineering have not really changed too much since the days of Archimedes: be it an automobile jack or a space station, it has to work, and work optimally at that. While the architect may use new methods, materials, and processes, the basic problems of human physique, circulation, planning, and scale are as true today as in the days of the Parthenon.

Industrial design, born at the beginning of the Great Depression, was at first quite properly a system that reduced manufacturing costs, made things easier to use, and improved the visual appearance of products along functional lines, to provide greater saleability on the chaotic market place of the thirties. But as the industrial designer has gained power over the design of more and more products, as designers have begun to function as long-range planners on upper managerial levels, members of the profession have lost integrity and responsibility and become purveyors of trivia, the tawdry and shoddy, the inventors of toys for adults and poor toys for children.

The opportunities for intelligent design today and tomorrow are greater than ever, for the world stands in need of wise reappraisal of its systems. America's economic stance abroad, the health and energy requirements of the world's people, global problems of water shortage, the need for mass housing, the combatting of disease, the waste of topsoil: long-range design planning can be part of the solution to these problems.

As we move further into our age of mass production, automation, conformity, and mass-media advertising, design has become ubiquitous. All of our means of communication, transportation, consumer goods, military hardware, furniture, packages, medical equipment, tools, utensils, etc., were designed for us. With a present world-wide need of 472 million individual family living units, it can be safely predicted that even "housing," still built individually by hand, will become a fully industrially designed mass-produced consumer product within a decade.

Admittedly, appliances, automobiles, sports equipment, all manner of communications equipment, structures, surgical tools, computers, teaching machines, space vehicles, and a myriad other objects, are well within the domain of the industrial designer. But if we equate man's basic needs with clothing, shelter, and tools, we see that exclusive fashion design and architecture are still entities separate from in-

dustrial design. But their first cousins, simple clothing and
mass shelter, are soon to be designed for mass production.

What is the contemporary architect, if not a master as
sembler of elements? At his elbow is *Sweet's Catalogue*, the
26 bound volumes that list most building components
panels, mechanical equipment, etc., that occupies an
honored place on the shelves of most architects' libraries
With its help, he fits together a puzzle called "house" or
"school" or whatever, by plugging in components—designed,
for the most part, by industrial designers and listed conveniently among the 10,000 entries in *Sweet's*. Where his
predecessors might have used marble fasciae, he substitutes
aluminum sandwich panels filled with polystyrene. (It is instructive to note that the handful of architects attempting to
design and build in a Wrightian, original, and innovating
manner, Bruce Goff, Paolo Soleri, Herb Greene, et al., have
actually collectively built 0.3 houses annually.) Quite
naturally some of the largest architectural offices, which
have a budget permitting the use of a 1401–1410 computer
set-up, merely feed all of *Sweet's* pages as well as the economic and environmental requirements of the job into the
computer, and let the computer "design" the building. With
endearing candor, some architects have taken pains to explain that "the computer does an excellent job."

At other times, as in the case of the new TWA Terminal
at Kennedy International Airport, the architect may create
what is no more than a three-dimensional trademark, an advertisement through which people are fed, but whose function it is to "create a corporate image" for the client, rather
than provide minimal comfort and facilities for passengers.
Having been trapped at the TWA Terminal during a fifteen-
hour power blackout, I can vouch for the inability of this
sculptural "environment" to process men, food, water, waste,
or luggage.

One of the difficulties with design-by-copying, design
through eclecticism, is that the handbooks, the style
manuals, and computer banks continuously obsolesce, go
out of style, and become old-fashioned and irrelevant to the
problem at hand. Furthermore, not only aesthetics is eliminated in designing via *Sweet's* and/or the computer: "The

Concert Hall and the Moonshot Syndrome," by William Snaith in his *Irresponsible Arts,* gives an excellent example of this.

The lacy mantles and Gothic minarets of Edward Durell Stone and Yamasaki are little more than latter-day extensions of the Chicago Fair of 1893. Frothy trifles, concocted to re-inject neo-romanticism into our prefabricated, prechewed, and predigested cityscape, can nonetheless be unconsciously revealing. For who could see Yamasaki's soaring Gothic arches at the Seattle Science Pavilion without realizing that here science was at last elevated through glib design clichés to the stature of religion? One almost expected Dr. Edward Teller to appear one Sunday morn, arrayed in laboratory vestments, and solemnly intone: "$E = mc^2$."

And what of Edward Stone's state legislature building in Raleigh, North Carolina? This semi-Oriental juke-box conveys none of the dignity which a building dedicated to legislation, deliberation, and law should expound. It does not declare its purpose (and hence many New Yorkers driving back from Miami try to "get a room for the night," mistaking it for a kissing cousin of the super-colossal motels they left that morning). What of an elderly woman, clutching $1 for a dog license, who must limp through 200 feet of marble halls to gain her objective? Walking down these endless marble vistas one always expects to find Mussolini at the other end, but alas, even this dubious promise is never fulfilled. Four identical pyramids crown this edifice; but while two quite properly house the Senate and House respectively, the others merely serve such varied activities as a restaurant and air-conditioning equipment, thereby again giving the lie to any formal integrity of structure.

If the need for some 472 million housing units around the world is to be met, surely the answer lies in mass production techniques and totally new concepts. The architect as supreme master builder, or the architect who defiles this fair land with gigantic sterile file cabinets, ready to be occupied by interchangeable people, or the speculative builder with his "boxes, little boxes" all are anachronisms in the seventies.

Buckminster Fuller, Jim Fitzgibbon of Synergetics, Inc.,

and a few other bold experimenters would shudder at the appellation "architect." But they are the type of designer whose comprehensive inventory of resources and needs in terms of global men, materials, tools, and processes will give us the industrially designed shelter of tomorrow.

When Moshe Safdie designed and built Habitat, an example of a radically new type of shelter, for the Montreal Exposition of 1967, he was among the first architect-planners who attempted to use a modular building system intelligently. Habitat has often been faulted for being both too expensive and too complex. In reality Habitat is probably the least expensive and at the same time most varied *system* that can be devised, and it is instructive to note that the Canadian Exposition Board made it impossible to build more than one third of the units. The strength of Habitat lies in the fact that once a large amount of money has been invested in basic building and handling equipment, the sys-

Modular Housing, shown first in the terraced houses and gardens of Habitat Montreal, then on the first site of Habitat Puerto Rico on San Patricio hill in Hato Rey, San Juan. Courtesy of M.I.T. Press and Tundra Books of Montreal. Photos by Jerry Spearman.

tem then begins to pay for itself as more units are built. For a fuller understanding of the Habitat system, see Safdie's two newer projects in Puerto Rico and Israel (cf.: R. Buckminster Fuller's *Nine Chains to the Moon*, page 37, and Moshe Safdie's book, *Beyond Habitat*).

Hemlines go up and hemlines go down, the pneumatic sweater girl of the forties changed to the shaggy, bulky look of the fifties, only to be replaced by the glistening, jack-booted, vinyl-clad teeny-bopper of the sixties; and necklines, we are promised, will plunge more deeply soon. Our young lady, window shopping in front of Paraphernalia, Inc., fully equipped with her "Frankly Fake Fur" miniskirt, electronic bra, black lace stockings and spike-heeled gold boots, has now finally emerged full-blown from the pages of Sacher-Masoch and Krafft-Ebing, read as arbiters of turned-on fashion. Men, gyrating from the "Bold" to the "Ivy-League" to the "Continental" to the "Carnaby Street" and later the "Virile" look, have also been at the mercy of fashion stylists. But here again, as in architecture, the industrial designer has entered the field of design for clothing through the back door, creating disposable work gloves (200 to a roll), ski boots, space suits, protective throw-away clothing for men handling radioactive isotopes, combat suits, and scuba gear. Lately, with the introduction of "breathing" and therefore really usable leather substitutes, much of the boot, belt, handbag, shoe, and luggage industry, too, is turning to the product designer for help. New techniques in vacuum form-ing, slush molding, gang turning, etc., make mass produc-tion design possible for products traditionally associated with handcrafted operations.

Thus tool, shelter, clothing, and breathable air and usable water are not only the job but also the responsibility of the industrial designer.

Mankind is unique among animals in its relationship to the environment. All other animals adapt themselves *auto-plastically* to a changing environment (by growing thicker fur in the winter, or evolving into a totally new species over a half-million-year cycle); only mankind transforms earth itself to suit its needs and wants (*alloplastically*). This job of form-giving and reshaping has become the designer's responsibility. A hundred years ago, if a new chair, carriage, kettle, or a pair of shoes was needed, the consumer went to the craftsman, stated his wants, and the article was made for him. Today the myriad objects of daily use are mass-produced to a utilitarian and aesthetic standard often completely unrelated to the consumer's need. At this point Madison Avenue must be brought in to make these objects desirable or even palatable to the mass consumer.

With products produced by the millions, mistakes too are multiplied a million times, and the smallest decision in design planning may have far-reaching consequences.

A simple example will suffice: Let's assume that the designers concerned with automotive styling in Detroit decide to move the car ashtray just 11 inches to the right, in order to establish greater "dashboard symmetry." And the results? *Twenty thousand Americans killed outright and another ninety thousand maimed on our highways within five years.* Almost an eighth of a million deaths and major accidents, caused by the driver being forced to reach just 11 inches further, thus diverting attention from the road for an extra 1/50 second. These figures are an extrapolation of the Vehicular Safety Study Program at Cornell University. It is interesting to note in this connection that, at the time of this writing, a General Motors executive has said that "GM bumpers offer one hundred per cent protection from all damage (and are therefore safe) *if the speed of the car does not exceed 2.8 mph*" (my italics). Meanwhile, the president of Toyota Motors is building a $445,000 shrine to "honor the souls of those killed in his cars." (Quoted in *Esquire* magazine, January 1971.)

Consider the home appliance field. Refrigerators are not designed, aesthetically or even physically, to fit in with the rest of the kitchen equipment. Rather, they are designed to stand out well against competing brands at the appliance store and scream for the consumer's attention.

Through wasting design talent on such trivia as mink-covered toilet seats, chrome-plated marmalade guards for toast, electronic fingernail-polish dryers, and baroque fly-swatters, a whole category of fetish objects for an abundant society has been created. I saw an advertisement extolling the virtues of diapers for parakeets. These delicate unmentionables (small, medium, large, and extra large) sell at $1 apiece. A long-distance call to the distributor provided me with the hair-raising information that 20,000 of these zany gadgets are sold each month.

In all things, it is appearance that counts, form rather than content. Let us go through the process of unwrapping a fountain pen we have just purchased as a gift. At first there is the bag provided by the store. Nestled in it is the package, cunningly wrapped in foil or heavily embossed paper. This has been tied with a fake velvet ribbon to which pre-tied bows are attached. The corners of the wrapping paper are secured with adhesive tape. Once we have removed this exterior wrapping, we come upon a simple gray cardboard box. Its only function is to protect the actual "presentation box" itself. The exterior of this little item is covered with a cheap leatherette that looks (somewhat) like Italian marble. Its shape conjures up the worst excesses of the Biedermeier style of Viennese cabinetry during the last and decadent stages of that lamentably long period.

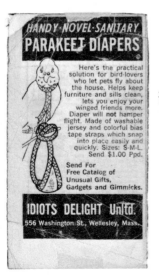

Advertisement for diapers to be used for parakeets. Author's collection.

When opened, the vistas thus revealed would gladden the heart of Evelyn Waugh's *Loved One*, for they match the interior appointments of a Hollywood-created luxury coffin to a nicety. Under the overhanging (fake) silk lining, and resting on a cushion of (phony) velveteen, the fountain pen is at last revealed in all its phalliform beauty. But wait, we are not yet done. For the fountain pen itself is only a further packaging job. A recent confection of this type (selling for 75 bucks) had its outer casing made not of *mere* silver, but out of silver obtained by melting down ancient "pieces of eight" recovered, one must assume, at great expense from some Spanish galleon fortuitously sunk near the Parker Pen factory 3 centuries ago. A (facsimile) map giving the location of the sunken ship and tastefully printed on (fake) parchment, was enclosed with each pen.

However, whatever the material of the pen-casing, within it we find a polyethylene ink-cartridge (cost, including ink: 3¢) connected to a nib. Thus, more than 80 per cent of the material consists of packaging, totaling (minimally) 90 per cent of the cost.

This example could be easily duplicated in almost any other area of consumer goods: the packaging of perfumes, whiskey gift decanters, games, toys, sporting goods, and the like. Designers develop these trivia professionally and are proud of the equally professional awards they receive for the fruits of such dedicated labour. Industry uses such "creative packaging"—this, it is useful to note, is also the name of a magazine addressed to designers—in order to sell goods that may be shabby, worthless, or just low in cost, at grossly inflated prices to consumers.

In the case of the silver pen cited above, the retail price of the silver pen in its package is approximately 145,000 per cent higher than the cost of the basic writing tool. We may say that inexpensive pens are, after all, available, and that the example mentioned merely illustrates "freedom of choice." But this "freedom of choice" is illusory, for the choice is open only to those to whom the difference between spending $75 or 19¢ is immaterial. In fact, a dangerous shift from primary use and need functions to associational areas has taken place here, since in most ways the 19¢ ball-point pen out-performs the $75 one. Additionally the tooling, advertising, marketing, and even the materials used in packaging represent such an exercise in futile waste-making that it cannot be countenanced in today's world. Well, what should package designers do?

As agricultural tools and implements, building components, and the like are shipped to the Third World, another need emerges in the field of package design: a way of imposing sequentially hierarchic assembly procedures on illiterate people, through the way in which the package surrounding the parts opens or unfolds. Anyone who has seen the huts hammered together out of flattened oil tins that provide shelter for millions of people throughout South America and South Africa must have asked himself why oil (and other raw materials) are not shipped in containers more suitable as building components. Re-use and different-use packaging is another major design challenge (for those package designers who feel like participating in rational work).

When we say that "the people's basic needs are satisfied," we are taking a narrow and parochial view. Right in our own country we are neglecting the needs of vast land-poor farm country, the multi-racial ghettos in our cities, the hard-core poor, and the mentally and physically retarded and disadvantaged. We also deliberately exclude 2,350,000,000 human beings in the so-called underdeveloped areas of the world. While it is true that the governments of the emerging nations of Africa, Asia, and the Americas are often engaged in the same frenzied, fetishistic rat-race to gain status symbols as nations as Americans are as individuals, the real needs of the broad masses of people remain unsatisfied. Substituting a national airline for our family Cadillac; building a cyclotron as we might build a family recreation room; installing a country-wide superhighway (with few cars and no service stations) as we might install air-conditioning; or using World Bank loans much as we might flash our charge-a-plates or Playboy keys: these examples of governmental status-seeking suffice.

Only a handful of American corporations and design offices are seriously engaged in facing up to challenges such as global minimal shelter needs, off-road vehicles for roadless terrain (84 per cent of the earth's land surface has no roads), new and compact teaching and training equipment geared to a society changing from a pre-literate one to a post-literate electronic one, and all on 1 HP per capita. The list is endless: power sources, basic medical, surgical and sanitation devices, food storage, communications, etc.

Several years ago I was approached by representatives of the United States Army and told of their practical problems concerning parts of the world (like India) where entire populations are illiterate and living on extremely low power levels. In many cases this means that the largest percentage of the population are unaware of even so basic a fact as their living in India. As they cannot read, and as there is neither enough power for radios, nor money for batteries, they are effectively cut off from all news and communication. I designed and developed a new type of communications device.

An unusually gifted graduating student, George Seegers, did all the electronic work and built the first prototype. It is a one-transistor radio, using no batteries or current, and designed specifically for the needs of developing countries. The unit consists of a used tin can. (As illustrated in this book, a used juice can is shown, but this is no master plan to dump American "junk" abroad: there is an abundance of used cans all over the world.) This can contains wax and a wick which will burn (just like a wind-protected candle) for about 24 hours. The rising heat is converted into enough energy (via a thermocouple) to operate an ear-plug speaker. The radio is, of course, non-directional. This means that it receives all stations simultaneously. But in emerging countries, this is of no importance: only one broadcast (carried by relay towers placed about 50 miles apart) is carried. Assuming that one person in each village listens to a "national news broadcast" for 5 minutes daily, the unit can be used for almost a year until the original paraffin wax is used up. At that time more wax, wood, paper, dried cow dung (which

Radio receiver designed for the Third World. It is made of a used juice can, and uses paraffin wax and a wick as power source. The rising heat is converted into enough energy to power this non-selective receiver. Once the wax is gone, it can be replaced by more wax, paper, dried cow dung, or anything else that will burn. Manufacturing costs, on a cottage industry basis: 9 cents. Designed by Victor Papanek and George Seeger at North Carolina State College.

The same radio as above, but decorated with colored felt cut-outs and sea shells by a user in Indonesia. The user can embellish the tin can radio to his own taste (Courtesy: UNESCO).

has been successfully used as a heat source for centuries in Asia, but for that matter anything else that burns will also work) will continue the unit in service. All the components: ear-plug speaker, hand-woven copper radial antenna, a "ground" wire terminating in a (used) nail, tunnel-diode, thermocouple, etc., are packed in the empty upper third of the can. The entire unit can be made for just below 9¢ (U.S.).

It is, of course, much more than a "clever little gadget." It is a fundamental communications device for pre-literate areas of the world. After it was tested successfully in the mountains of North Carolina (an area where only *one* broadcast is easily received), the device was demonstrated to the Army. They were shocked. "What if a Communist," they asked, "gets to the microphone?" The question is meaningless, since the most important business before us is to make information of all kinds freely accessible to the people. After further developmental work, I gave the radio to UNESCO. UNESCO in turn is seeing to it that it is distributed to villages in Indonesia. No one, neither the designer, nor UNESCO, nor any manufacturer, makes any profit or percentages out of this device since it is manufactured as a "cottage industry" product.

In 1966 I showed color slides of the radio at the *Hochschule für Gestaltung*, at Ulm in Germany. It was interesting to me that nearly all the professors walked out (in protest against the radio's "ugliness" and its lack of "formal" design), *but all the students stayed.* Of course, the radio *is* ugly. But there is a reason for this ugliness. It would have been simple to paint it ("gray," as the people at Ulm suggested). But painting it would have been wrong. For one thing, it would have raised the price of each unit by maybe one twentieth of a penny each, which is a great deal of money when millions of radios are built. Secondly, and much more importantly, I feel that I have no right to make aesthetic or "good taste" decisions that will affect millions of people in Indonesia, who are members of a different culture.

The people in Indonesia have taken to decorating their

tin can radios by pasting pieces of colored felt or paper, pieces of glass, and shells on the outside and making patterns of small holes toward the upper edge of the can. In this way it has been possible to by-pass "good taste," design directly for the needs of the people, and "build in" a chance for the people to make the design truly their own.

This is a new way of making design both more participatory and more responsive to people in the Third World.

It is true that during the fifties some large design offices, such as Chapman & Yamasaki of Chicago, Joe Carreiro of Philadelphia, and others, performed design development work in underdeveloped countries at the request of the State Department. But their work was largely in the area of helping young nations to design and manufacture objects that would appeal to the American consumer. In other words, they did not design for the needs of the people of Israel, Ecuador, Turkey, Mexico, etc., but rather for the fancied wants of American purchasers.

Our townscape as well bears the stamp of irresponsible design. Look through the train window as you approach New York, Chicago, Detroit, Los Angeles. Observe the miles of anonymous tenements, the dingy, twisted streets full of cooped-up, unhappy children. Pick your way carefully through the filth and litter that mark our downtowns or walk past the monotonous ranch houses of suburbia where myriad picture windows grin their empty invitation, their tele-viscous promise. Breathe the cancer-inducing exhaust of factory and car, watch the strontium-90 enriched snow, listen to the idiot roar of the subway, the squealing brakes. And in the ghastly glare of the neon signs, under the spiky TV aerials, remember, this is our custom-designed environment.

How has the profession responded to this situation? Designers wield power over all this, power to change, modify, eliminate, or evolve totally new patterns. Have we educated our clients, our sales force, the public? Have designers attempted to stand for integrity and a better way? Have we tried to push massively forward, not only in the market place but in the affections of needy people of the world?

Listen in on a few imaginary conversations in our design offices:

> "Boy, wrap another two inches of chrome around that rear fender!"
>
> "Somehow, Charlie, the No. 6ps red seems to communicate freshness of tobacco more directly."
>
> "Let's call it the 'Conquistador' and give people a chance for personal identification with the sabre-matic shift control!"
>
> "Jesus, Harry, if we can just get them to PRINT the instant coffee right onto the paper cup, all they'll need is hot water!"
>
> "Say, how about roll-on cheese?"
>
> "Squeeze-bottle martinis?"
>
> "Do-it-yourself shish-kebab kits, with disposable phenolic swords?"
>
> "Charge-a-plate divorces."
>
> "An aluminum coffin, communicating 'nearness-to-God' (non-denominational) through a two-toned anodized finish?"
>
> "A line of life-sized polyethylene Lolitas in a range of four skin shades and six hair colors?"
>
> "Remember, Bill, the corporate image should reflect that our H-bombs are always PROTECTIVE!"

These imaginary conversations are quite authentic: this is the way designers talk in many offices and schools, and this is also the way in which new products often originate. I might add that industrial designers actually enjoy reading such dialogues. One proof of this authenticity is that, of the eleven idiocies listed above, all but two—charge-a-plate divorces and "protective" H-bombs—are now on the market.

You may wonder if this isn't just a hysterical outburst, directed towards some of the phonier aspects of the profession? Are there no dedicated designers working away at jobs that are socially constructive? The truth is that, of all the articles in the professional magazines, of all the talks at design conferences, few indeed have dealt with professional responsibilities, responsibility going beyond immediate market need. The latter-day witch doctors of market analysis, motivation research, and subliminal advertising have made dedication to meaningful problem-solving rare and difficult.

The philosophy of most industrial designers today is based on five myths. By examining these, we may come to understand the *real* underlying problems:

1. *The Myth of Mass Production:* In 1966, 16 million easy chairs were produced in the United States. But if we divide this number by the 2,000 manufacturers of such chairs we find that, averaging it out, only 8,000 chairs could have been produced by each manufacturer. If we further realize that each manufacturer has, on the average, 10 different models in his line, this reduces our number to only 800 chairs of one kind. If we now add the fact that furniture manufacturers' lines change twice a year, in time for the spring and fall market showings, we will find that, on the average, only 400 units of any given chair were produced. This means that the designer, far from working for 200 million people (the market he is trained to think about) has, on the average, worked for 1/5,000th of 1 per cent of the population. Let's contrast this with the fact that in backward and underdeveloped areas of the world there exists a present need for close to 2 billion inexpensive, basic seating units.

2. *The Myth of Obsolescence:* Ever since the end of World War II, an increasing number of responsible people in the very top levels of management and government have voiced the myth that, by designing things to wear out and be thrown away, the wheels of our economy can be kept turning *ad infinitum* and *ad nauseam*. This is patent nonsense. One of the healthiest companies in the United States is the Polaroid Corporation. Even though new models of Polaroid Land cameras have replaced earlier models over the years, none of the old cameras are obsolete because the corporation is careful to continue manufacturing film and accessories for them. The German Volkswagen has moved into a leading position in sup-

plying the transportation needs of the world by carefully refraining from major style changes or cosmetic jobs. The Zippo cigarette lighter sells at a far better rate than all other domestic lighters combined, even though (or could it be because?) the manufacturer guarantees to repair or replace its case and/or guts for life. (In fact, it is outsold only by foreign, point-for-point copies.) There is ironic justice in that. For it was in 1931 that George Grant Blaisdell, a non-smoking American, noticed that some of his friends carried wind-proof, dependable, Austrian cigarette lighters which sold in chain stores for twelve cents. He tried importing them directly and selling them at $1 apiece, but finding that the public was unwilling to pay that much during the Depression, he quit temporarily. He waited for the patent of the Austrian model to expire and began producing it in 1935 and offering it for sale with a life-time guarantee. The Zippo lighter has moved from an item made on $260 worth of second-hand tools, in a $10 room in Brooklyn, to a production level of 3 million units per year. Since many of our products are beginning to be obsolesced technologically anyhow, the question of forced obsolescence becomes redundant and, in terms of raw materials, a dangerous doctrine indeed.

3. *The Myth of the People's "Wants":* Never in recent times have the so-called "wants" of people been investigated as thoroughly by psychiatrists, psychologists, motivation researchers, social scientists, and other miscellaneous tame experts, as in the case of the ill-fated "Edsel." That mistake cost $350 million and led one comedian to quip that the mistake "was being handled by the Ford Fund."

"The people want chrome, they like tailfins," except the Volkswagen and the Fiat exploded that idea thoroughly. So thoroughly, in fact, that Detroit had to start producing compact cars a few years ago when foreign imports began to seriously affect American sales figures. As soon as foreign imports began to drop off, compact cars were again advertised as "the biggest, longest, lowest, most luxurious of them all." This stylistic extravaganza has now once more raised the number of small European cars coming into the country.

Where are they now?

Return with us now to those wondrous days of yesteryear.

It's 1949 and automobiles are getting longer, lower and wilder.

Massive bumpers are a big hit. Fins are in. And everyone's promising to "keep in style with the times."

But then, times changed.

Massive bumpers and fins went out. So did every car shown above, except the VW.

You see, back in '49, when all those other guys were worrying about how to improve the way their cars looked, we were worrying about how to improve the way ours worked.

And you know what? 2,200 improvements later, we still worry about the same thing.

A Comparison of Automobiles of 1949. Advertisement by Volkswagen of America, Inc.

4. *The Myth of the Designer's Lack of Control:* We
are told that it is "all the fault of the front office, the
sales department, market research," etc. But of 150
mail-order, impulse-buying items foisted on the public
during the last few years, a significantly large number
were first conceived, invented, planned, patented, and
produced by members of the design profession. These
products include such inspiring items as:

"Mink-Fer," a tube of deodorized mink droppings sold
at $1.95 each as a Christmas fertilizer for "the plant
that has everything."

A $1,595 electronic computer for practicing golf
swings. This tidy item makes it possible to play golf
in the bathroom or cellar without ever having to go
outdoors at all.

A $39.95 electronic clip-on gadget that attaches to
the front of the automobile and flashes the message
"You're welcome!" when the electronic traffic light in a
pay-it-yourself highway toll booth lights up to say
"Thank you."

5. *The Myth That "Quality No Longer Counts":* While
Americans have for years bought German, and later,
Japanese cameras, Europeans now line up to buy Polaroid Land cameras and equipment. American "Head"
skis are outselling Scandinavian, Swiss, Austrian, and
German skis around the world. Sales of Schlumbohm's
Chemex coffee maker are diminished only somewhat
by a recent German copy of it. The United States Army
Universal Jeep designed by Willys in 1943 (since modified, and sold by American Motors) is still one of the
most desirable multi-purpose vehicles on earth; its
only major foreign competitors are the British Land
Rover and the Japanese Toyota Land Cruiser, both up-
dated and improved versions of the Jeep.
    The one thing which these and some other American
products that still command world leadership hold in
common is a basically new approach to a problem,
excellent design, and the highest possible quality.

Something can be learned from these five myths. It is a fact that the designer often has greater control over his work than he believes he does, that quality, basic new concepts, and mass production could mean designing for the majority of the world's people, rather than for a small domestic market. Designing for the people's *needs* rather than for their *wants,* or artificially created wants, is the only meaningful direction now.

Having isolated some of the problems, we must ask what can be done about them. At present there are several fields in which little or no design work is being done. These are areas that are, by their very nature, highly profitable to manufacturer and designer alike. They are areas that promote the social good that can be inherent in design. All that is needed is a selling job, and that is nothing new to the industrial design profession.

It is possible to outline briefly a number of important areas in which the discipline of industrial design is virtually unknown:

1. *Design for Underdeveloped Areas:* Over two billion people stand in need of some of the most basic tools and implements.

Today more oil lamps and other kinds are needed globally than before the discovery of electricity because there are more people without electric power alive today than the entire global population in Thomas Edison's day. In spite of new techniques, materials, and processes, no radically new oil lamp (or for that matter, primitive light source) has been developed for 106 years.

Eighty-four per cent of the world's land surface is completely roadless terrain. Often epidemics sweep through an area: nurses, doctors, and medicine may be only 100 kilometers away, but there is no way of getting through. Regional disasters, starvation, and water shortages also develop frequently: again there seems to be no good way of getting through. Helicopters work, but are far beyond the monies and expertise available in many regions of the Third World. Beginning in 1962, a graduate class and I developed an off-road vehicle that might be useful for such emergencies. We asked that it fulfill the following performance characteristics:

a. The vehicle would operate on ice, snow, mud, montane forests, broken terrain, sand, certain kinds of quicksand, swamps, etc.

b. The vehicle would cross lakes, streams, and small rivers.

c. It would climb 45° inclines and transverse 40° inclines.

d. It would carry a driver and 6 people, or a driver and a 1,000-pound load, or a driver and 4 stretcher cases; finally it would be possible for the driver to walk next to the vehicle, steering it with an external tiller, and thus carry more load.

e. The vehicle could also remain stationary and, with a rear-power takeoff, drill for water, drill for

Mock-ups and working models of two vehicles designed and built under the author's direction at Konstfackskolan in Stockholm, Sweden. These vehicles were explorations in transporting materials over rough terrain by muscle power alone. One of them (designed by James Hennessey and Tillman Fuchs) is a proposal for an inner-city run-about and shopping vehicle. It will carry two people and 200 pounds (Courtesy: *Form* magazine).

oil, irrigate the land, fell trees, or work simple lathes, saws, and other power tools.

By inventing and testing a completely new material, "Fibergrass" (*sic*)—using conventional chemical fiberglass catalysts, but substituting dried native grasses, hand-aligned, for the expensive fiberglass mats—we were able to reduce costs. Over 150 species of native grasses from all parts of the world were tested. By also attacking the manufacturing logistics, it was possible to reduce costs still more. Various technocratic centers were established: heavy metal work was to be done in the United Arab Republic, Katanga, Bangalore (India), and Brazil. Electronic ignitions were to be made in Israel, Japan, Puerto Rico, and Liberia. Precision metal work and the power train were to be done in the Chinese Democratic Republic, Indonesia, Ecuador, and Zambia. The Fibergrass body would be made by users all over the world. Several prototypes were built (and are illustrated elsewhere in this book), and it was possible to offer the vehicle to UNESCO at a unit price of less than $150.

But this is where responsible design must begin to operate. The vehicle worked fine, and in fact, UNESCO told us that close to 10 million vehicles might be needed initially. But the net result of going ahead with this would have meant introducing 10 million internal combustion engines (and consequently, pollution)

Off-Road Vehicle, discontinued for ecological reasons, designed by student team under the author's direction, School of Design, North Carolina State College, 1964.

into hitherto undefiled areas of the world. So we have shelved the off-road vehicle project until a better power source is available.

(Historical footnote: as I do not believe in patents, photographs of our vehicle were published in a 1964 issue of *Industrial Design* magazine. Since then, more than 25 brands of vehicles, priced between $1,200 and $2,000, have been offered to wealthy sportsmen, fishermen, and (as "fun vehicles") to the youth culture. These vehicles pollute, destroy, and create incredible noise problems in wilderness areas. The destructive ecological impact of the "snowmobile" is detailed in Chapter Ten.)

At this point, as a result of our concern for pollution and together with a group of Swedish students at *Konstfackskolan* in Stockholm, we began exploring muscle-powered vehicles. The Republic of North Vietnam moves 500-kilogram loads into the southern part of that country by pushing these loads along the Ho Chi Minh trail on bicycles. The system works and is effective. However, bicycles were never designed to be used in just this maner. One of our student teams was able to design a new type of vehicle, made of bicycle parts, that would be more effective. The new vehicle is specifically designed for pushing heavy loads; it is also designed to be pushed easily uphill through the use of a "gear-pod" (which can be reversed for different ratios, or removed entirely). The vehicle will also carry stretchers, and, because it has a bicycle seat, it

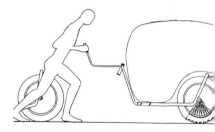

can be ridden. Several of these vehicles plug into each
other to form a short train. (Photographs and sketches
are reproduced elsewhere.)

These drawings show that the muscle-powered vehicle
can be plugged together into a short train. It also comes
apart, and the geared power pod is reversible so that the
vehicle can be pushed uphill under heavy loads. It can
also carry stretchers or with the power pod removed, be
used like a wheelbarrow. Designed under the author's
direction by a student team in Sweden, it could be used
in underdeveloped areas to propel heavy loads, similar
to the loads pushed on bicycles along the Ho Chi Minh
Trail in North Vietnam. (Photos by Reijo Rüster. Courtesy
Form magazine.)

As bicycles are needed a[s] transportation devices in the Third World, this luggage carrier was designed to flip down and be used as a temporary power source when needed. Its construction is within the scope of the most modest village technology. Designed by Michael Crotty and Jim Rothrock, as students at Purdue University.

When students suggested the use of old bicycles or bicycle parts, they regretfully had to be told that old bicycles also make good transportation devices and that parts are always needed for replacement or repair. (They may have been influenced somewhat negatively by the fact that a design student recently won first prize in the Alcoa Design Award Program by designing a power source intended for Third World use, made of brand-new aluminum bicycle parts.)

So we designed a new luggage carrier for the millions of old bicycles all over the world. It is simple and can be constructed in a village. It will carry more payload. But it will also fold down in 30 seconds (see illustration) and then can be used in its primary capacity for generating electricity, irrigation, felling trees, running a lathe, digging wells, pumping for oil, etc. After this use, the bicycle can be folded up again and returned to *its* primary function: a transportation device. Except that it now has a better luggage carrier.

A Swedish student built a full-size sketch model of a vehicle that is powered by the arm muscles and can go uphill. This in turn led us at Purdue University to design an entire generation of muscle-powered vehicles that are specifically designed to provide remedial exer-

cise for handicapped children and adults. Photographs of these vehicles are shown in this book.

For shopping and short-distance hauling of bulky packages, I have said that a simple 3-wheel bicycle with a storage compartment would do extremely well. To help the rider in going uphill, an "assist" motor that is electrically powered and rechargeable might be provided. I see on page 41 of the Abercrombie & Fitch Company's Christmas catalogue for 1970 that they offer such a vehicle for sale. In fact, if necessary, it can attain speeds of 40 mph. If the rider chooses, he can (and should) pedal, of course. However, the A & F vehicle retails for $650. I have seen it demonstrated. There is no need for the price to exceed $90. Unfortunately, one of New York's most prestigious stores has bestowed the aura of "Upper Westchester Status Object" on it, so the price now reflects this philosophy.

2. *Design of Teaching and Training Devices for the Retarded, the Handicapped, the Disabled, and the Disadvantaged:* Cerebral palsy, poliomyelitis, myasthenia gravis, mongoloid cretinism, and many other crippling diseases and accidents affect one tenth of the American public and their families (20 million people) and approximately 400 million people around the world. Yet the design of prosthetic devices, wheel-chairs, and

Tricycle for adults with battery power-assist. $650 each. (Courtesy Abercrombie & Fitch Co.)

other invalid gear is by and large still on a Stone Age level. One of the traditional contributions of industrial design, cost reduction, could be made here. At every Rexall or Walgreen drugstore it is possible to buy a Japanese transistor radio for as little as $3.98 (including import duties and transportation costs). Yet as mentioned previously, pocket-amplifier-type hearing aids sell at prices between $147 and $600 and involve circuitry, amplification needs, and shroud design not radically more sophisticated than the $3.98 radio.

Hydraulically powered and pressure-operated power-assists are badly in need of innovation and design.

Robert Senn's hydrotherapeutic exercising water float is designed in such a manner that it cannot be tipped. There are no straps or other restraint devices that would make a child feel trapped or limited in its motions. At present hydrotherapy usually consists of having the child strapped to a rope attached to a horizontal ceiling track. In Robert Senn's vehicle all such restraints are absent. Nonetheless, his surfboard-like device is safer (it will absorb edge-loading of up to 200

water vehicle designed for hydrotherapy of handicapped children. igned by Robert Senn, as a graduate student at Purdue University.

pounds), and the therapist can move in much more closely to the child. Later, I explain further ideas we have developed in this field.

3. *Design for Medicine, Surgery, Dentistry, and Hospital Equipment:* Only recently has there been responsible design development of operating tables. Most medical instruments, especially in neurosurgery, are unbelievably crude, badly designed, very expensive, and operate with all the precision of a snow-shovel. Thus a drill for osteoplastic craniotomies (basically a brace and bit in stainless steel) costs $125 and does not work as sensitively as a carpenter's brace and bit

System of drills and saws for osteoplastic craniotomies. Design copyrig and designed by C. Collins Pippin, North Carolina State College.

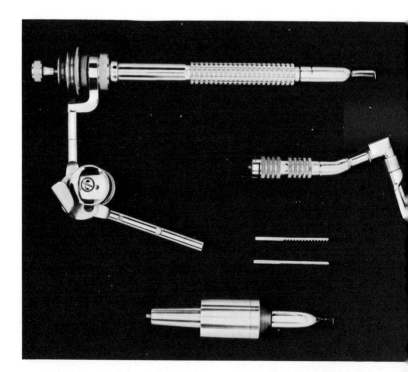

available for $5.98 at Sears Roebuck. Skull saws have not changed in design since predynastic times in Egypt. As mentioned before, one of my graduating students was able to develop a radically new power-driven drill and saw for osteoplastic craniotomies, which, in wet labs devoted to experiments with animals, revolutionized the entire neurophysiological field.

The cost of health care for the "poor" is rising astronomically. Regardless of who it is that absorbs these costs in the long run, the fact is that a great deal of the high expense can be attributed directly to bad design.

From time to time, illustrations of new biomedical equipment appear. Almost invariably these are "hi-style modern" cabinets, in nine delicious decorating colors, surrounding the same old machine. Hospital beds, maternity delivery tables, and an entire host of ancillary equipment are almost without exception needlessly expensive, badly designed, and cumbersome.

4. *Design for Experimental Research:* In thousands of laboratories doing research, most of the equipment is antiquated, crude, jury-rigged, and high in cost. Animal immobilization devices, stereo-encephalotomies, and the whole range of stereotactic instruments need intelligent design reappraisal.

With million-dollar grants from the National Institutes of Health, the National Research Foundation, and many other governmental and private foundations showering largesse upon university research departments, there has been a steady and steep climb in the price of laboratory instruments. In one case in the area of bio-electronics a simple meter lists for 8,000 per cent above the retail price of all its components, and assembly time for the unit has been estimated at less than 2 hours. A company in New York manufactures a simple electric lab timer. This unit can be purchased by amateur photographers for $8.98. The identical unit can also be purchased by research laboratories for $172.50. A hand mixer is offered to the housewife in two versions: white enamel finish ($13.98), or stainless steel ($15.98). For laboratory use, the same unit by the same manufacturer is listed at $115.00 in white enamel and $239.50 in stainless steel. Certainly this is an area in which honest design, value engineering techniques, and cost reduction could play an important part. It might even be possible to manufacture and sell laboratory apparatus at an *honest* profit, for a change.

5. *Systems Design for Sustaining Human Life Under Marginal Conditions:* The design of total environments to maintain men and machines is becoming increasingly important. As mankind moves into jungles, the Arctic, and the Antarctic, new kinds of environmental design are needed. But even more marginal survival conditions will be brought into play as sub-oceanic settlements and experimental stations on asteroids and other planets begin to make their appearance. Design for survival in space has already become important.

The pollution of water and air and the problems of our sprawling city-smears also make a re-examination of environmental systems design necessary.

6. *Design for Breakthrough Concepts:* Many of our products have by now reached a dead end in terms of further development. Designers merely *add* more and more extra gadgets rather than re-analyzing the basic problems and trying to evolve totally new answers. Automatic dishwashers, in the First and Second Worlds, waste billions of gallons of water each year (in the face of a world-wide water shortage), even though newer systems such as ultrasonics for "separating-dirt-from-objects" are well within the state of the art. The rethinking of "dishwashing" as a system might not only make it easier to clean dishes, but would also help solve one of the basic survival problems of humanity today: water conservation. Our toilets, as mentioned previously, also waste water.

There exists at present a world-wide need for nearly

350 million television sets to be used largely for educational purposes. Groups from the United States, bidding for some of these contracts of African and Asiatic markets, are being turned down for one very simple reason: line definition. As discussed in Chapter Four, American television sets have a definition of 525 lines to the inch (our penalty for having been first), most of Europe uses a 625-line definition, France 819, and the Soviet Union 625. With some justice, many European, Asiatic, African, and South American people feel that the eyes of their children might be spoiled by exposing them to the inferior line definition of American television sets. And here again we have made the fatal error in continuing to design for our own needs (our own transmitting facilities are set up for a 525-line definition). It is curious that the leaders of the TV industry do not realize that in having, say, a 1,000-line-definition set designed, they would not only capture the world's markets, they might also gain from a "slop-over" effect. For with so many technically better sets on hand, Americans would then have reason enough to change our own antiquated equipment.

Messrs. Alexander Salosin and Viktor Prokhorov of Donetsk in the Soviet Union have designed a thimble-like insert for men's smoking pipes. It is a gadget intended for people whose vocal cords are weak or semi-paralyzed, and it contains a generator sending out sound oscillations of 80–90 cycles per second. This makes it possible for people with paralyzed vocal cords to make themselves understood. This too is a break-through approach that has been suggested to American manufacturers, only to be laughed out of the office as not having enough saleability.

Humidity control in homes and hospitals is important and sometimes can become critical. In many regions of the United States humidity levels are such that humidifiers and de-humidifiers find a ready market. These gadgets are costly, ugly, and ecologically extraordinarily wasteful of water and electricity.

In researching this problem for a manufacturer, Robert Senn, I, and some others were able to develop a theoretical humidifier/de-humidifier that would have no moving parts, use no liquids, pumps, or electricity. We decided to use deliquescent crystals. By combining a mix of deliquescent crystals, anti-bacteriological crystals, etc., we were able to develop a theoretical surface finish that would store 12 atoms of water to each crystal atom and release it again when humidity was unusually low. This material could then be sprayed onto a wall, woven into a wall-hanging, or whatever, and do away with the drain on electric power as well as with the noise pollution and expense of present systems.

Here again the problems are endless, and not enough solutions are coming from our own designers.

These are six possible directions in which the design profession not only can but must go if it is to do a worthwhile job. So far the designers have neither realized the challenge nor responded to it. So far the action of the profession has been comparable to what would happen if all medical doctors were to forsake general practice and surgery, and concentrate exclusively on dermatology and cosmetics.

9

THE TREE OF
KNOWLEDGE: BIONICS
The Use of Biological
Prototypes in the Design of
Man-Made Systems

> *A bird is an instrument working according to mathematical law, which instrument it is within the capacity of man to reproduce with all its movements.*
>
> LEONARDO DA VINCI

ONE HANDBOOK that has not yet gone out of style, and predictably never will, is the handbook of nature. Here, in the totality of biological and biochemical systems, the problems mankind faces have already been met and solved, and through analogues, met and solved optimally.

The ideal solution to any problem in design is always to achieve "the mostest with the leastest," or to use George K. Zipf's happy phrase, "the principle of least effort."

By now a definition of the word bionics is probably in order: bionics means "the use of biological prototypes for the design of man-made synthetic systems." To put it in simpler language: to study basic principles in nature and emerge with applications of principles and processes to the needs of mankind.

Dr. Edward T. Hall states in *The Hidden Dimension* that "man and his environment participate in molding each other. Man is now in the position of actually creating the total world in which he lives, or what the ethologists refer to as his biotype. In creating this world he is actually determining *what kind of an organism he will be.*"

Indeed, the problems confronting the designer have become truly frightening in terms of complexity and urgency of solution. With 472 million basic living shelters needed on earth, a number enlarging at the rate of 16 million per month, the architect, contractor, and speculative builder have plainly abrogated their capability to house the world's teeming billions.

Even the smallest problem in the area of product design will illustrate that a great deal more than a designer with a modicum of "good taste" is needed: several years ago a new low-cost plow was designed, built, and distributed in areas of Southeast Asia that until then had used a forked stick weighted down by a rock to till the soil. After a few years it was discovered that the plows were not in use and were, in effect, rusting away. According to the religious beliefs of the inhabitants, metal makes the soil "sick," and offends the Earth-mother. I was able to recommend that the plows be dipped in a plastic compound similar to Nylon 60. And as the people were not offended by the technology of plastics, the new plows were accepted and usefully employed.

The point of this anecdote is that the use of a cross-disciplinary design team, including anthropologists, engineers, biologists, psychologists, sociologists, etc., would have prevented the original mis-design. Shifting back to a more sophisticated design level, one can see that if William Snaith's "Concert Hall," cited earlier, had had the services of consultant musicologists and sociologists as part of the design team, both the "dry, acrid sound" and the shortage of seats could have been avoided.

At present industrial and environmental designers are the logical foci in any design team. Their logical status as key synthesist in a design situation is not because they are superior beings, better informed, or necessarily more creative but rather because they assume their status as comprehensive synthesist by the *default* of all other disciplines. For in this age in America, education in all the other areas is a matter of increasing *vertical specialization*. Only in industrial and environmental design is education *horizontally cross-disciplinary*.

While the designer in any team situation may know far less psychology than the psychologist, far less economics than the economist, and very little about, say, electrical engineering, he will invariably bring a greater understanding of psychology to the design process than that possessed by the electrical engineer. By default, he will be the bridge.

The basic tenets on which this chapter is based are:

1. That the design of products and environments, on or off earth, must be accomplished through interdisciplinary teams, until such time as sleep-learning telepathy or the extension of the human life span make it possible and practical for the designer-planner to be conversant with all the parameters of the problem.

2. That biology, bionics, and related fields offer the greatest area for creative new insight by the designer.

3. That the design of a single product unrelated to its sociological, psychological, cityscape surroundings is no longer possible or desirable. Therefore, the designer must find analogues, using not only bionics but biological systems design approaches culled from the fields of ecology and ethology.

Man has always looked to nature and derived ideas from the workings of nature, but in the past this has been achieved on a very simple level. Technological design problems have, however, become increasingly complex during the last 100 years, and, with the proliferation of technology in our society, mankind has become more and more alienated from direct contacts with his biological surroundings.

Designers and artists especially have looked to nature, but their viewpoints have often been clouded by a romantic longing for the re-establishment of some sort of primeval Eden, a desire to get back to "basics" and escape the depersonaliz-ing power of the machine, or by a sentimental mystique about "closeness to the soil."

And interestingly enough, virtually nothing has been writ-ten in the area of bionics. Heinrich Hertel's *Structure, Form and Movement* (1963), Lucien Gerardin's *Bionics*, and E. E. Bernard's *Biological Prototypes and Man-Made Systems* (1963) are about all that has appeared in book form. For the most part, these three books and the various reports on bionics prepared by the armed services concern themselves with man-computer-control relationships only and deal with the interface between cybernetics and neurophysiology. There have been a few articles in the *Saturday Evening Post, Mechanics Illustrated,* and *Industrial Design,* but these have been largely over-simplified popularizations.

Of course, through history there have been the exceptional designers. *"A bird is an instrument working according to mathematical law, which instrument it is within the capacity of man to reproduce with all its movements,"* said Leonardo da Vinci in 1511. Fire, the lever and fulcrum, early tools and weapons—all these were invented by man observing natural processes, with the wheel possibly the only exception to this rule. And even here Dr. Thomasias presents a closely reasoned argument for the wheel having been derived from observation of a log rolling down an inclined plane.

During the last 100 years, and especially since the end of World War II, scientists have begun looking into the biologi-cal sciences in a search for answers in problem-solving areas and have managed to find new breakthroughs that are of enormous importance to today's technology. An important difference between early man and today's designs must be made at this point: While we may consider the first hammer an extension of the fist, the first rake a type of claw, and we pityingly smile at the attempt made by Icarus to fix bird wings to himself and fly into the sun, today bionics is con-cerned not so much with the *form of parts* or the *shape of things,* but rather, with the possibilities of examining *how* nature makes things happen, *the interrelation of parts, the existence of systems.*

Thus a psychologist, shown the diagram of a control me-chanism for a recent apparatus enabling a blind man to read by scanning letter forms and transforming them into tones, immediately recognized it as the so-called fourth layer of the visual cortex, the part of the brain responsible for Gestalt vision.

As far back as the early calculating machines, scientists recognized a similarity between the machine's function and the function of the human nervous system. With the advent of vacuum tubes, the similarity becomes even more startling. It is for this reason that one of the more active areas today in bionics lies in the field of computer design. Here, insights from computers to human brains and from human brains to computers have been gained during the last decade and a half. Professor Norbert Wiener at M.I.T. worked with psychologists, physiologists, and neurophysiologists to attempt to learn more about the human brain through the construction of computers whereas Dr. Heinz von Foerster, in work with Professor W. Ross Ashby and Dr. W. Grey Walter at the University of Illinois, has gained insight into the way computers ought to be constructed through his research on the design of the human brain.

W. Grey Walter, the British physiologist mentioned above, managed to evolve simple electronic machines that responded positively to light as a stimulus source. In other words, these machines will head for the nearest light source: a research finding much indebted to the study of the photophiliac behavior of the common moth.

Rattlesnakes are known to biologists as pit vipers because of the two pits located in the snake's snout midway between nostrils and eyes. These pits contain temperature-sensing organs so delicate that they can detect temperature changes of 1/1,000 of a degree. This might be the difference, for instance, between a sunbaked stone and a motionless rabbit. A similar principle has been used by Philco and General Electric in the design of the sidewinder missile, a heat-seeking air-to-air missile which homes in on the exhaust of jet aircraft.

Bats find their way in the dark through an echo location

method: they emit a high-pitched sound which bounces off objects in their path, is picked up by their sensitive ears, and thus establishes an unencumbered flight path for them. Much the same principles are used in radar and sonar. Sonar uses audible sound waves; radar uses ultra-high-frequency waves.

One excellent example of a bionic design investigation is a remarkably accurate speed indicator for airplanes that was developed using the same principle found in beetles' eyes. It was discovered that certain beetles compute their air speed prior to landing by watching moving objects on the ground. A study of the sense organs of these beetles has given us a present aircraft speed indicator for measuring the time elapsed between its passage over two known points on the ground and translating it into speed.

Dr. Ralph Redemske, a specialist in the area of bionics now working for Servomechanism, Inc. in Santa Barbara, California, has recently plated an ordinary honeybee with a thin coating of aluminum. Using the standard black background, this enabled him to make photographs (which were less fuzzy than a bee) of every detail of its complex structure. From this work, engineers may some day create mechanical eyes modeled from those of bees.

One of the most interesting animals that holds many different promises of design solutions is the bottle-nosed dolphin (*Tursiops truncatus*). The dolphin uses a radar, sonar-like navigational system which does not depend on hearing. In common with other whales, it ripples its external skin surface, utilizing this effect for navigation and increased swimming speed.

The ground effects caused by a helicopter flying in a stationary position at a distance of less than 50 feet above ground level have puzzled aircraft engineers for over a decade. Only recently, through a study of the dragonfly, are those causes beginning to be understood.

The question of energy input versus output is another interesting one: two examples in this area are the South American fruit bat, or flying fox, and the male of a South American beetle called *Acroncinus longimanus*. In the case of the fruit bat, its truly gigantic wing-spread and great power require a comparatively small energy input. The incredibly long front legs of the South American beetle utilize even less energy input and derive great payload power.

I found the input-output disparity of the beetles to be a challenging problem. Eventually I was able to dissect several of these beetles. What is at work is a energy amplification system utilizing fluid. It is a measure of my naïveté that I immediately assumed, gleefully, that I had succeeded in a theoretical breakthrough. And it is a fact that if I had dissected these beetles some forty years ago (at the tender age of five) I would today be known as the "Father of Fluidics." In all seriousness though, there is a point hidden in this rambling anecdote: unknown to me, fluidics existed. But it is abundantly clear that there is an infinite number of biological principles—like fluidics—lying around, waiting to be discovered.

The major emphasis in industrial and environmental design, however, will certainly lie in the ethological and ecological approach to systems, processes, and environments. In recent years, when industrial designers talked about "Total Design," they referred to two things. First, that the design of, say, a steam iron might lead also to the design of the logo, the manufacturer's letterhead, the point-of-sale displayer for the iron, the package, and possibly even some control over the merchandising of the product. At other times, "Total Design" meant in-plant work: the design of the handling machinery for manufacturing the steam iron, safety de-

Acroninus longimanus, male specimen showing elongated front legs. Author's collection.

vices, traffic patterns within the plant, etc. "Total Design" in the future will mean seeing the steam iron as well as its plant and promotional gimmicks merely as links in a lengthy biomorphic phylogenetic chain reaching back to heated rocks and stove-irons, and forward to the final extinction of the phylum "steam iron" by mass introduction of "perma-pressed" and "stay-press" fabrics.

If the industrial revolution gave us a *mechanical* era (a static technology of movable parts), if the last 60 years have given us a *technological* era (a dynamic technology of functioning parts), then we are now emerging into a *biomorphic* era (an evolving technology permissive of imitations).

We have been taught that "the machine is an extension of a man's hand." But even this no longer holds. For 5,000 years, a brickmaker was capable of making 500 bricks a day. Technology has made it possible for one man, with the right kind of back-up machinery, as described in Chapter Four, to make 500,000 bricks a day. But biomorphic change obsolesces both the man and the bricks: we now extrude building skin surface, i.e., sandwich panels that include heating, lighting, cooling, and other service circuits.

While Mr. Robert McNamara, the former United States Secretary of Defense, may often have found himself ambivalent regarding our involvement in Southeast Asia, it may be salutary to think that the same Mr. McNamara, while with the Ford Motor Company, shared the then prevailing automotive infatuation with tailfins, hood ornaments, and other small and neo-Freudian ephemera. Had the American automotive industry brought its production know-how to something like the self-generating styrofoam domes developed by Dow International, some 250 million shelters might have "grown" in Southeast Asia by now, and the sociological pressures leading to civil wars and American involvement might never have happened.

This total chain of design can probably be best explained anecdotally. According to an old German ecological legend, old maids are responsible for the growth of the British Empire. It seems that bumblebees are the only insects with tongues long enough to pollinate red clover. There are more bumblebees near English towns since there are fewer field-mice near towns, and fieldmice have a disconcerting habit of eating the combs and larvae of bumblebees. The fieldmouse population is kept down by the high correlation between old maids and cats in England. Consequently, there is more red clover which forms a staple food for cattle. Cattle are made into bully beef, a staple food for the Royal British Navy. The Navy made Great Britain into a world power and gave her an empire lasting some 300 years. Thus, there is a high correlation between old maids and the way the British Empire was built.

Recently, while visiting some friends in Canada, I bought some turtle food for them in a little quarter-ounce can for 50¢. The contents were labeled as "dried flies and fly ashes." In other words, dead flies are worth $32 a pound. Let's design an electronic fly-catcher. If we can sell it to the consumer for $10 including sufficient packaging for 10 pounds of dead flies and promise to buy the first 10 pounds back from him for $10, our consumer will break even and we shall show a net profit on the order of $310.

But there is more than a joke in ecological links. Consider the fact that the absorption of 10,000 pounds of Radiolaria establishes 1,000 pounds of plankton, that 1,000 pounds of plankton establish 100 pounds of small marine animals, these in turn create 10 pounds of fish, and it takes 10 pounds of fish to put one pound of muscle tissue on a human being. The frictional losses in the system are simply staggering. With 168,000 species of insects in North America there is 6 to 8 times as much insect protein living in a 40-acre field as beef protein represented by grazing cattle thereon. Actually, we do eat flies; it's just that we process them through grass, cows, and milk first.

It may be argued that the "average" industrial designer or design engineer concerned with research and development lacks a sufficient background in the biological sciences to utilize biology meaningfully as an inspirational source of design. If we attempt to define the word "bionics" in its narrowest sense, that is on a cybernetic or neurophysiological level, this may be true. But all around us are manifestations in nature of rather primitive structures that have never been properly investigated, exploited, or used by designers, biological schemes that bear investigation and are accessible to anyone free for a walk on a Sunday afternoon.

Take seeds, for instance. A simple maple seed (*Aceraceae saccharum*), when released from just a few feet off the ground, will fall in a very definite spiral pattern. This method of air-to-ground delivery has so far never been applied in any significant way. In Chapter Five, I described the use of artificial maple seeds as part of a system for soil erosion control. One of the more interesting applications of the maple seed's flight characteristics discovered by a design student was a new method of extinguishing forest fires, or rather, getting fire-extinguishing modules into inaccessible parts. An artificial maple seed some 8⅔ inches long was constructed out of inexpensive, ultra-lightweight plastic. The seed portion contained a fire-extinguishing powder. Experimentation and investigation showed that when maple seeds were released above a fire, they would naturally be caught up in the thermal up-drafts above the flames. If, on the other hand, the seeds were *forced* below the up-draft area and into the semi-vacuum below, their flight pattern would re-establish itself, and they would, in fact, fly towards the hottest part of the fire. To return to the plastic maple seeds. Thousands of these encased in time-sacks would be dropped from airplanes. The sack would rip open once it had plunged under its own gravity to below the up-draft area. Then thousands of plastic, expendable maple seeds would circle towards the hottest part of the fire, and here, their casing consumed by the flames, the fire extinguisher would be released. This is by no means a way to put out forest fires. It is, however, a way of getting at canyons and other areas that are normally not accessible from the ground or to smoke-jumpers.

Reforestation of the extreme northern tundra areas of Alaska, Canada, Lapland, and the Soviet Union, as well as the restocking of these areas with fish, could be achieved through water-soluble maple seeds that contained seed spores or fish eggs. Naturally these artificial maple seeds could also easily include nutrient solutions, serve as thermo-protectors, or carry fertilizer.

The random spreading of almost any material can be achieved through artificial maple seeds; tolerances are re-assuringly broad: I have constructed artificial maple seeds that performed optimally with a wing-spread up to 46 inches. At the other end of the scale, maple seeds only ¼ inch long can be operational.

The seed of the white ash (*Fraxinus americana*) has characteristics very similar to those of the maple seed. In still air the seed falls almost straight down, spinning in a tight area. In a strong wind the seed will travel horizontally or, because of its lightness, climb for a while while spinning rapidly. If the seed's mass were concentrated into a small solid sphere, it would fall much faster, due to the decreased surface area, which would decrease the frictional drag act-ing on the body. However, if the seed were a hollow sphere of the same mass and with the same surface drag, but did not spin, it would fall still faster. Thus we see that the spinning actually helps to slow the seed's descent. This is due to the fact that in spinning, the seed uses energy that would otherwise contribute to its rate of descent.

Basswood seeds (*Tilia americana*) are distinguishable by their unusual flight pattern. The "wings" force a spinning motion as the seeds slowly descend, drifting with the wind in spite of the (comparatively) great weight of the double-seed which sticks out from the wing part on bifurcated ex-tenders.

The flight characteristics of all of these spiraling seeds have not yet been studied sufficiently. The spiraling behavior of such seeds, artificially created in media other than air (water, oil, gasoline, etc.) or in near-vacuum or different gravity situations, may also prove a rich source of design concepts. We shall concern ourselves with the behavior of only one other seed in this group.

The falling Ailanthus seed (*Ailanthus altissima*) falls spinning rapidly about its longitudinal axis, making one complete revolution while descending about one quarter of its length. The physical geometry of this seed can be ap-

ONE REVOLUTION

IN STILL AIR, THE ASH SAMARA FALLS ALMOST STRAIGHT DOWN, SPINNING IN A TIGHT SPIRAL. THE ASH DESCENDS APPROXIMATELY 1/4 OF ITS LENGTH DURING EACH REVOLUTION OF SPIN. IN STRONG WINDS THE SEED WILL TRAVEL HORIZONTALLY OR EVEN CLIMB WHILE SPINNING RAPIDLY. IN SPINNING, THE ASH USES ENERGY THAT OTHERWISE WOULD BE CONTRIBUTING TO ITS RATE OF DESCENT.

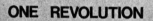

ash samara descent

Four examples of research into aerodynamic behavior of seeds. Graduate team research under the author's supervision by Robert Toering, John K. Miller, and Jolan Truan, as students at Purdue University.

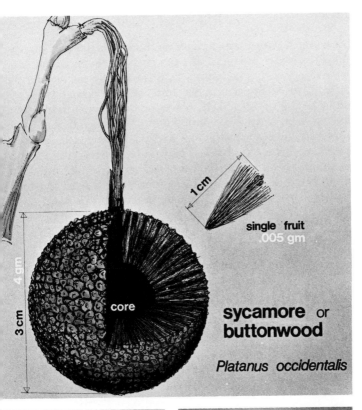

1 cm

single fruit
.005 gm

core

4 gm

3 cm

sycamore or **buttonwood**

Platanus occidentalis

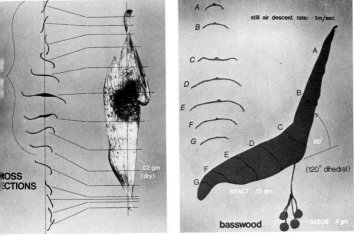

A
B
C
D
E
F
G

CROSS
SECTIONS

.02 gm
(dry)

still air descent rate: 1m/sec

A

B

C

60°

D

E

(120° dihedral)

F

G

BRACT .07 gm

basswood

SEEDS .4 gm

proximated with twisted paper as shown. In the first simulation, the twists produced at each end are equal, which only very rarely occurs in nature. In this case the seed descends, in still air, along a straight line, at approximately a 45° angle to the horizontal. If, however, the twists are unequal, as shown in the second simulation, the seed follows a path that combines a spiral action and a screwing, axial spin at the same time. The twisted end pulls air from the vicinity of the tip of the seed in towards the center of the seed. This produces a high-pressure area around and under the seed, which slows its descent. When the twists are equal, they

An example of research into the aerodynamic properties of seeds in flight. This example is the Ailanthus seed. Graduate team research by John K. Miller and Jolan Truan, Purdue University.

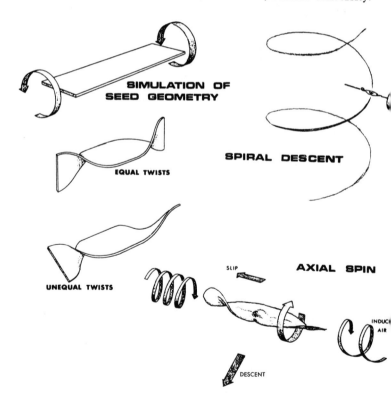

SIMULATION OF
SEED GEOMETRY

EQUAL TWISTS

SPIRAL DESCENT

UNEQUAL TWISTS

SLIP   AXIAL SPIN

INDUCE
AIR

DESCENT

both push the same amount of air towards the center, producing no unequal forces. However, when the twists are unequal, the end with the greater twist will pull more air, producing a lower pressure in the vicinity of that end. Therefore, the seed is being acted upon by unequal forces. The seed tends to slip axially towards the lower pressure area. Thus, instead of following a straight line, the seed descends in a spiral path. The combination of axial spin, slip, and spiral descent gives a very slow and almost random flight pattern to each seed.

Seeds of the wild onion (*Allium cernuum*) and the salsify plant follow flight patterns of an entirely different configuration. The wild onion seed is a delicate structure of radiating, lacy, "umbrella"-type formations. Dozens of these form a spider-web-like ball around the plant's central hub. As shown, the umbrellas are closely interconnected and inverted slightly. When released, the delicate filaments flatten out and lose their convexity. They fall like tiny parachutes, only at a much slower rate. Because, unlike parachutes, they possess a flat, disc-shape top consisting of scores of finely interlaced hairs, their rate of fall, directionality, etc., may be applicable to uses far different from those of conventional parachutes. Their lacy mantle also would foil radar detection.

Anchorage, grappling, and hook-closures are another seed characteristic. The common cocklebur (*Xanthium canadense*) will cling to an animal's fur or, for that matter, to a man's trousers when he walks across a field in autumn. The specific hooking action has been adapted to "Velcro" nylon closure strips. Here a female surface of tiny loops and a male surface of tiny hooks are biaxially oriented. When pressed together, they can be pulled apart only in one direction and resist parting along all other axes. As mentioned above, this principle has been used in the basic Velcro closure for clothing, but has also recently been adopted for the upper-arm bands used in determining blood pressure; also recently American astronauts wearing the male part on the soles of their feet walked on drop-cloths consisting of the female part that were lashed to the exterior of a space capsule, thus making walking in a null gravity situation possible.

Non-electric toothbrushes have a tendency to destroy the enamel of the teeth, not to get into inaccessible areas, and also to become prime incubators for germs. After brushing one's teeth in the morning the (still moist) toothbrush is hung up in the warmest, most humid spot of the house: the bathroom. Thus, the bristles of the toothbrush become a wonderful area for growing cultures of microbes and bacilli, which are then carefully smeared all over one's mouth at night. A new type of toothbrush, exploiting the cocklebur-Velcro closure, was therefore developed. The consumer would be given a free handle covered with a female surface. He would purchase hundreds of tiny cushions or pillows consisting of the male surface and filled with tooth powder. In the morning one could wet one of the pillows, literally just lay it on the handle, and brush away. Afterwards the little cushion could be peeled off and thrown away, to be replaced by a new one when next needed.

Explosive-force seeds—seeds that, because of the interior construction of the seed pod, are hurled 20 feet or more—will provide another useful area of research. Specifically the seed of a small berry, *Hubus arcticus,* growing only in the Lapland section of Finland, would repay investigation.

The very simplest growth characteristics of almost any plant may provide the solution to imaginative design problems. Thus the growth of an ordinary green pea may become instructive. If the pea is permitted to "go to seed," at one growth stage a string at the back of the peapod ceases to grow. As the rest of the pod continues to enlarge, within a few days it very slowly opens, and the pea seeds are slowly raised above the level of the pod. A manufacturer of suppositories for children was persuaded to adopt this concept in his package. Hitherto, each suppository was separately wrapped in silver foil, a dozen or so to a box. The parent, when unwrapping them, would soon find three quarters of the glycerine substance under his fingernails, with the suppository, of course, now de-sterilized. By creating a package of polyethylene that had been purposely miscast, this problem was solved. The package had been so cast that the "memory" of the plastic was its "open state." The suppositories, sterilized, but with all need for wrapping gone, would now be inserted, and a high-impact styrene closure would slide over the top. The small polyethylene package would not be under tension. Its purposeful miscasting would act like the string on the back of the peapod. When the styrene top was gently slid off, the package would open very slowly and the suppositories gently be forced up and out. Closing the package would merely mean squeezing it gently together (thus forcing the remaining suppositories back down) and sliding the styrene retainer-top back on.

Nothing has been said about insulating, heat-storing, protection from cold, and many other properties possessed by seeds.

An equally large area for bionic design investigation lies in the field of botanic architecture, such as growth patterns, cells, and growth rate of bamboo shoots, the architecture of a rose, various stem configurations of plants, and the properties of mushrooms, algae, fungi, and lichen. Regarding

This package was bionically derived from a pea pod. Author's design.

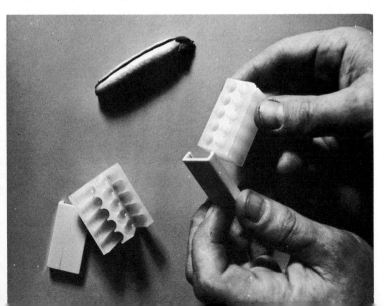

this last item let me discuss an example (with indebtedness to William J. J. Gordon):

In facing the problem of repainting the interior of buildings, the cost of paint, labor, and depreciation has to be considered. It is an obvious fact that a freshly painted room may look beautiful for several days or weeks, but the slow, inexorable process of deterioration starts. Let's try (still with Bill Gordon) to isolate the problem. Paint is a substance which "looks good" when first applied to the wall but looks more and more dilapidated as time goes on. The problem then: Is it possible to find a substitute which, when first applied to the wall, may look unpleasant but which will be self-improving and self-maintaining? The answer is not far away. Lichen (a symbiotic growth relationship between algae and fungi) comes in nature in a selection of some 118 "delicious decorator colors." We could theoretically select the lichen of our color choice, spray it on a wall together with a

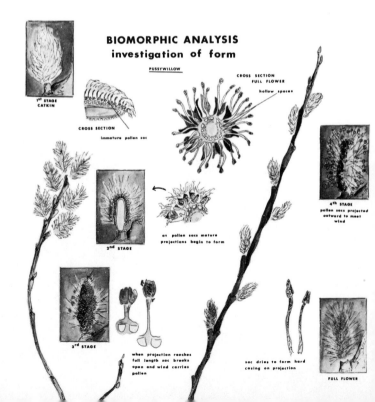

BIOMORPHIC ANALYSIS
investigation of form

PUSSYWILLOW

1ˢᵗ STAGE CATKIN

CROSS SECTION
immature pollen sac

CROSS SECTION
FULL FLOWER
hollow spaces

4ᵗʰ STAGE
pollen sacs projected
outward to meet
wind

as pollen sacs mature
projections begin to form

2ⁿᵈ STAGE

3ʳᵈ STAGE

when projection reaches
full length sac breaks
open and wind carries
pollen

sac dries to form hard
casing on projection

FULL FLOWER

nutrient solution and just sit back and relax. Obviously the wall may be a splotchy-looking mess at first, but as the lichen begins to grow, an even color will result. Unfortunately, the designer may be obliged to speculate as to whether people can be motivated to enjoy shaggy walls. But a serious application is possible. Nearly all lichens grow a height of approximately one and a half inches, and as they are not affected by such temperature extremes as 30° below zero or 125° F., one direct use would be to plant them, instead of grass, in the center median of the New York Thruway. Because mowing costs of the New York Thruway Authority at present amount to some $2.5 million annually, this would be a great saving indeed. Furthermore, color-coding might be brought into play: the Berkshire cut-off could be planted in, say, blue, and the Ohio cut-off in red.

The growth patterns of the pussy willow have led a student to evolve a seed-planting tool that could be used in the

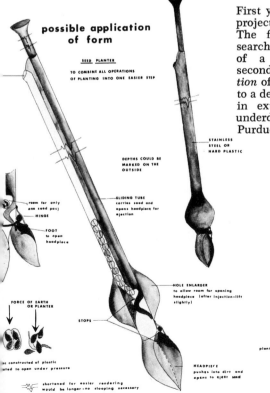

possible application of form

SEED PLANTER

TO COMBINE ALL OPERATIONS OF PLANTING INTO ONE EASIER STEP

DEPTHS COULD BE MARKED ON THE OUTSIDE

SLIDING TUBE carries seed and opens headpiece for ejection

room for only one seed pac)

HINGE

FOOT to open headpiece

STAINLESS STEEL OR HARD PLASTIC

HOLE ENLARGER to allow room for opening headpiece (after injection-lift slightly)

FORCE OF EARTH OR PLANTER

STOPS

ac constructed of plastic ated to open under pressure

HEADPIECE pushes into dirt and opens to eject seed

planter and seed pac designed from pollenation principle

shortened for easier rendering would be longer—no stooping necessary

First year freshman student project in bionic research. The first plate shows research into the configuration of a pussy willow. The second plate is an *application* of these basic principles to a device for planting seeds in extremely hard soil in underdeveloped countries. Purdue University.

underdeveloped areas of the world where the soil is poor and hard. This simple hand tool utilizing a basic bionic principle could be of great use specifically in Central India, Shansi, and Sinkiang, as well as the Mongolian People's Republic. Furthermore, the tool is simple and maintenance-free, so that it might be used by relatively unsophisticated people in the Central Kalahari Desert.

Turning to an entirely different area now, let us see how we can exploit the world of crystallography. If asked to fill completely two-space with polygons of the same type and size, there are only three ways in which the job can be done: a grid of equilateral triangles, squares, or hexagons. Even though the number of polygons is infinite, we cannot derive a complete "space-fill" from them. Octagons, for instance, require small squares for fill-ins; with pentagons the job is impossible.

If we attempt the same thing in three-space, there are again only a very limited number of possible solutions. We can use bricks, which are, after all, a type of square-ended prism. For the same reason we can also utilize equilaterally-triangular-ended prisms or hexagonal prisms. Proceeding with any of these three symptoms, we have merely built a two-dimensional construct in space. Using any of these three grid patterns, we can make a wall as high or long as we wish; its depth, however, will be that of one brick. True integration in three dimensions has not taken place.

If we derive our shape from the field of crystallography and semi-regular polyhedra, we will find that there is one shape, and one shape only, that makes a stable, fully three-dimensionally-integrated space grid possible: the tetrakaidecahedron.

Tetra (four) kai (and) deca (ten) hedron: a fourteen-sided polyhedron consisting of eight hexagonal and six square faces. A number of these will cluster easily in space because their angles of incidence and adherence make the job very easy. If we examine one of the shapes, we will find that it is rounder than a cube but squarer than a sphere. This may lead us to feel that it can resist pressure (either from without or from within) better than a cube but not as well as a sphere. True, but only for a single solid. If we cluster a series of spheres of the same size (balloons for instance) like a cluster of grapes and subject them to equal and steady pressure, for instance, by submerging in water, we will find that little pressure areas (in the form of convex, spherical, triangular pyramids) build up between our balloons. If pressure is permitted to build up more at some point, the balloons will collapse into their most stable shape: a cluster of tetrakaidecahedra. The tetrakaidecahedron, in fact, is the idealized shape of the human fat cell, as well as many other basic cellular structures in nature.

Here again a series of tetrakaidecahedra were handed to students for design exploitation. Many completely new design solutions resulted. By building huge tetrakaidecahedral cells 38 feet in diameter, it was possible to construct a sub-oceanic shelter area for men and materials that might

Tetrakaidecahedra: archimedian solids that close-pack in three-dimensional space.

be used for sub-ocean mining, oil drilling, etc. Each cell consisted of three floor levels; a cluster of between 30 and 90 of these cells would constitute a sub-oceanic station.

By reducing the diameter of the cells to ⅛ inch, a new type of radiator for an automobile was evolved, exhibiting more surface areas and containing more water.

A folding, semi-permanent vacation house, sleeping 20, could be, in its knock-down stage, transported in a standard VW camper.

By again building the tetrakaidecahedra to a 38-foot module, a central tower could be erected, 11 units or 418 feet tall. Twenty-eight more units of the same size could then be attached in spiral form, surrounding the central core. With each unit being tri-level, the result is a luxury apartment building. The central core tower would carry stairways, air-conditioning conduits, elevators, water, heat, and electricity. In addition, any given central core unit (also being tri-level) would house bathrooms, kitchens, and other service rooms, each level housing rooms for the closest cantilevered spiral unit. The 3 floors of the exterior spiral unit could be given over to living, entertainment, and sleeping areas, with the hexagonal roof of each acting as a combination heliport and garden. Because of the ease with which further omnidirectional units can be "plugged in" or, for that matter, "un-plugged," each tetrakaidecahedron that is part of the exterior spiral could easily be air-lifted and plugged in other core units in different locations of the world. It is obvious that the same construction might also serve as a constantly contracting or expanding grain silo, etc.

When the first visual model of this structure was removed from its base, I attached a line to it and dragged it through water (much like taking your dog for a walk on a leash). It has excellent motion characteristics in water. This opens the way for constructing huge hollow tetrakaidecahedra out of ice (reinforced with algae), pumping them full of crude oil, and towing a string of these spiral clusters across the Atlantic by submarines, thus eliminating the need for tankers.

The answer most elegant technologically, however, lies in the field of space stations. Assume that a basic cluster of

tetrakaidecahedra (each tri-level and 38 feet in diameter)*
numbering 48 single units were to be put in a locked orbit
150 miles above earth. This unit could house a labor force
of 300 men. If we now place further single cells in orbit we
will find that (because of the many angles of incidence and
adherence, mentioned above) 300 workers can attach another 50 units in a 24-hour work period. At this point the
station (which incidentally would provide enough centrif-
ugal spin to give a semblance of earth gravity from a central
atomic pile) will house 600 people. After 2 days of work it
will house 1,200 workers, 9,600 at the end of 5 days, 307,-
200 at the end of 10 days and 9,830,400 at the end of 15
days. In other words, it would be possible to absorb the entire populations of, say, Finland and Austria, or else all of
Greater New York, in 2 weeks, *with all of these people in
tri-level structures.* Now give the whole construct a push,
and when it arrives at, say, Mars, Alpha Centauri II, or Wolf
359, it will be possible to decant people *and their homes,*
establishing a city at the same speed as people can be
landed.

All this experimentation was done during 1954–1959.
Now other exploitations of crystalline forms are possible.
William Katavolos of New York has suggested that cities
could be "grown." With recent breakthroughs in Russian
crystallography and our increased abilities to grow large
hollow crystals, it may be possible, before long, to "seed" an
entire city and move in when it is fully grown.

The snub rhomboicosadodecahedron, consisting of 80
equilateral triangles and 12 pentagons, lends itself quite
naturally to the erection of dome structures. While these
domes bear a generic resemblance to Buckminster Fuller's
geodesic domes, they are in fact simpler to erect, for all sides
are straight and of equal length, and all angles identical.

The articulation of a snake's skeleton finds its application
in a variable curve ruler by Keuffel & Esser Company. Again

---

* The 38-foot-diameter module has been established as a
"principle-of-least-effort" structure. In other words, it exploits
sandwich-skin panels to their utmost. Larger constructs are
feasible, of course, but only with a sharp rise in cost.

it may be worthwhile to point out that in this, as in all other cases, bionic design application never means copying by establishing a visual analogue. Rather, it means searching out the basic, underlying organic principle and then finding an application.

A whole group of beetles: *Propomacrus bimucronatus, Euchirus longimanus, Chalcosoma atlas* & *Forma colossus, Dynastes hyyllus* & *centaurus, Dynastes hercules* & *Granti horn* & *Neptunus quensel,* the *Megasomae (elephans, anubis, mars, gyas),* and the *Goliathi* (especially *Goliathus Goliathus drury, atlas, regius klug, cacius, albosignatus, meleagris* and the *Fornasinius fornasinii* & *russus* as well as the *Meoynorrhinse* & *Melagorrhinae,* the *Macrodontiae* and especially the *Acrocinus longimanus* L. [males only]) have "front-end" handling mechanisms that are startling in their variety and challenging in their complexity. None of these has ever been intelligently exploited.

Even in mentioning some random thoughts on shell structures and sea shells, regeneration, exoskeletal structures, various propulsion systems in fish, the swimming behavior of snakes, "free" soaring in flying fish, we barely scratch the surface of a few of the areas that will yield to bionic design development.

John Teal, professor of human ecology at the University of Alaska, is attacking the problem of the musk ox. With 48 chromosomes, the musk ox is actually not an ox but related to the goat and antelope. Also, there is no musk. The fur of the musk ox is actually better than wool in terms of moisture-shedding and heat-retaining properties. John Teal has set for himself the rather unusual job of domesticating musk oxen and eventually giving the results of his studies to Eskimo tribes and Laplanders all across the northern tundra belt of the world. A completely new human ecology and social pattern should emerge among these deprived northern people, based on a spinning and weaving trade. At present, the normal ratio of the musk ox birth-rate is three females to one male, a problem that can be eliminated through multiple-birth-hormone injections. One of the reasons why Dr. Teal's work is so unusual is that no musk ox has been domesticated by man in nearly 6,000 years.

THE TREE OF KNOWLEDGE: BIONICS   209

Mere odd speculations about the possible future domesti-
cation of microbes may open entire new vistas in design-
planning, bionically, for medical applications.

In some areas of design, almost direct translations of
natural phenomena can be used. In Düsseldorf in 1940, a
gigantic vertical turning machine was constructed by having
the interior "sperm" machine build the rest of the machine
around itself.

Greater London, with a population close to that of New
York City and with an unbelievably primitive and leaky
water supply system, nonetheless uses only one fourth as
much water as is consumed in New York. The reason is a
biomorphic one. D'Arcy Wentworth Thompson quotes Roux
in formulating the following empirical rules for the branch-
ing of arteries and leaf venation:

>   1. If an artery bifurcates into two equal branches,
> these branches come off at equal angles to the main stem.
>   2. If one of the branches be smaller than the other,
> then the main branch or continuation of the original
> artery makes with the latter a smaller angle than does
> the smaller or lateral branch. And
>   3. All branches which are so small that they scarcely
> seem to weaken or diminish the main stem come off
> from it at a large angle, from seventy to ninety degrees.

The water supply of London was laid out according to the
above rules and, in spite of marginal losses, represents a
biologically stable system.

In some fields "by-passing" characteristics are beginning
to make their appearance. Sonic thesia, a system used recently in dental work, equips the patient with a pair of
stereo earphones through which he listens to pre-recorded
music. A third stream broadcasts a continuous screaming or
wailing sound which the patient has to tune out continuously with a pain control. The patient becomes so task-
oriented that little or no pain is felt because nerve endings
and pain receptors are being by-passed.

In a similar way, a Bell Telephone Systems proposal to
give people a standard billing per month, which would permit
unlimited station-to-station direct-dialing calls anywhere on
the continent, makes a great deal of sense, because right

now billing procedures for individual long-distance call
cost more than the phone calls themselves. When an opera
tor broke in on your local nickel telephone call in the thir
ties, it was still good business to do so. Today, with full
automated equipment, communication satellites, and a lac
of operators, this no longer makes sense even on a long-dis
tance basis. Surely world-wide direct dialing will "by-pass"
present political and national boundaries.

Turning far beyond even this rudimentary "by-passing"
through supra-national bodies, opens the even wider field o
environmental design. Here, of course, a rather limited ver
sion of cross-disciplinary approaches has been long pre
dicted. Design teams have included many diverse elements
architects, city planners, landscape architects, regiona
planners, and the occasional sociologist.

Nonetheless, it is precisely in the area of environmenta
design that bionic approaches and biological insights gleaned
from the most recent research in ecology and ethology wil
be most valuable. As we create the regional smear reaching
from Kansas City to St. Louis to Chicago to Cleveland to
Erie to Buffalo, we also participate in creating inhabitants
of prisons, slums, redeveloped slums, suburbs and exurbs
mental institutions, and $35,000 condominiums. The subtle
interaction of all these marginal types, as well as their inter
action with the dominant culture, has yet to be studied
interpreted, and understood.

But even more frightening are the recent studies performed
with animals under stress conditions and extreme crowding.
Fatty degeneration of the heart and liver; brain hemorrhage;
hypertension; atherosclerosis with its attendant effects of
stroke and heart attack; adrenal deterioration; cancers and
other malignant growths; eye strain; glaucoma and tra-
choma; extreme apathy, lethargy, and social non-partici-
pation; high abortion rates; failure of mothers to tend their
young; extreme promiscuity among the barely pubescent;
a rise in homosexuality, lesbianism, and the emergence of
a new sexual sub-type given to impressive and colorful, but
superficial, displays of his virility, though in reality ex-
tremely passive or even asexual; this may sound like a list of
what some people think of as moral decay or the ailments

of modern urbanized people, but it is not. The symptoms listed above have been observed in such widely divergent animals as Minnesota jackrabbits, Sika deer, Norway rats, and several species of birds. The common denominator has always been stress syndromes caused by overcrowding. Similar behavior patterns have also been observed among concentration camp inmates, prisoners, etc. It has caused Dr. John Calhoun of the National Institute of Mental Health to coin the accurate and lethal phrase "pathological togetherness."

Up to now environmental planning has sublimely disregarded all this.

Some sixty-five years ago, Henri Van de Velde—an early designer-artist-craftsman and theoretician of aesthetics—called the light bulb "revolutionary." The revolutionary aspect, however, was not a function of the light bulb; rather, it was a function of Mr. van de Velde, for he attempted to apply *traditionally aesthetic* standards of "Formal Necessity" and "Logic of Purpose" to an *industrial* product. The same kind of fuzzy nineteenth-century thinking can be seen when the Museum of Modern Art exhibits the tangled color-coded guts of a computer as "Twentieth-Century Good Design."

Industrial and environmental design are one of the few fields in which the schools are ideologically in the forefront of the profession. In spite of some anti-intellectuals in the design field who have just done "a good job on a luggage handle" or "really communicated no-mow, no-grow consumer satisfaction through Sassygrass" (an outdoor nylon carpet in red, brown, blue, or green), design, as the most powerful shaping tool yet developed by mankind to manipulate himself and his environment, will go on. The professional society meetings that endlessly and fruitlessly attempt to define industrial design might take another look at the sciences. Electricity, after all, is never defined but is described as a function, its value being expressed in terms of relations—the relation between voltage and amperage, for instance. Industrial and environmental design, too, can be expressed only as a function; its value, for instance, being expressed in terms of relations: the relation between human ability and human need.

10

CONSPICUOUS

CONSUMPTIVES:

DESIGN AND THE

ENVIRONMENT

Pollution, Crowding,

Starvation, and the

Designed Environment

*Nature has let us down, God seems
to have left the receiver off the
hook, and time is running out. . . .*
ARTHUR KOESTLER

IF DESIGN is ecologically responsive, then it is also
revolutionary. All systems—private capitalist, state socialist, and mixed economies—are built on the assumption that
we must buy more, consume more, waste more, throw
away more, and consequently destroy Liferaft Earth. If
design is to be ecologically responsible, it must be independent of concern for the gross national product (no matter
how gross that may be). Over and over I want to stress that

n pollution, the designer is more heavily implicated than most people. By now the garbage explosion has outdistanced the population explosion, and as Professor E. Roy Tinney, the director of the State of Washington's Water Research Center, has remarked, "We have not run out of water. We have simply run out of new streams to pollute." The strength of our chemicals has grown to the point where in mid-July of 1969, one single 200-pound sack of the German pesticide "Thiodan" that accidentally fell off a barge on the Rhine was able to kill more than 75,000 tons of fish in Germany, Holland, Switzerland, Austria, Liechtenstein, Belgium, and France *and to stop a new fish population from forming for a time period that has been estimated at 4 years.* The automobile is held accountable now for more than 60 per cent of all air pollution generated in the United States, and for an increasing amount in other Western nations.

Scientists are beginning to realize that jet aircraft pollute the upper atmosphere (as there is no "washing effect" at extremely high altitudes), and so pollutants from aircraft will circle the earth many times before settling through gravitation. Dr. Alfred Hulstrunck (Atmospheric Research Center, State University of New York) comments: "If transportation continues to grow the way it's going, it's possible the next generation may never see the sun." Were this to happen (and there is a good chance that it will by 1990), this might not necessarily spell global darkness. Instead, the "hothouse effect" might take over. Transparent to sunlight, but opaque to the earth's radiation, a blanket of moisture and carbon dioxide might raise the surface temperatures of the earth enough to melt the polar icecaps. This would, at the very least, raise sea levels by 300 feet (shrinking habitable land by 64 per cent). But in all likelihood, the sudden weight shift might spin the earth off its axis.

We could continue with these examples at a faster rate than the speed at which this can be read. And so far we have dealt with "neutral" interventions, that is, things that have just seemed to happen, rather than results, or foreseeable results, of malign intent.

We need spend little time on the fact that malign intervention can pollute and kill. The 5,000 sheep killed by the accidental release of a United States Army nerve gas in Skull Valley, Utah, on March 21, 1968, bear mute witness to the dangers of chemical warfare.

But what happens when man's intentions are *benign* from the start? Through the building of the Aswan Dam, Egypt attempted to make a quick transition from 6,000 years of agricultural history to twentieth-century technology. One of the largest structures of its kind, the Aswan Dam project was specifically designed to provide a multitude of socio-economic benefits. There would be, minimally, a 25 per cent increase in cultivated land, and electrical output would double. Unfortunately, things have not worked out like that. Lake Nasser (part of the Aswan development) retains most of the silt on which the rich Nile Delta farmland depends. The dam also impounds essential natural minerals, needed by the ecological chain of marine life in the delta. Since Aswan first started to regulate the flow of the river in 1964, Egypt has suffered a loss of $35 million to its native sardine industry; as of the spring of 1969, there are reports that the delta shrimp fishery is also declining.

Professor Thayer Scudder of Cal Tech has reported similar results in the wake of damming up the Zambezi River in southern Africa. The designers of the dam had predicted that the loss of flooded farmland would be offset by increasing fishery resources. In reality, the fish catch diminished immediately after the dam was completed, and, soon after, the lake shore bred hordes of tsetse flies, which infected native livestock and nearly aborted cattle production.

But we have learned nothing from these lessons. At the time I was writing this, engineers were designing the largest dam systems in the history of mankind for two of the world's longest rivers: the Mekong and the Amazon. The Hudson Institute's proposal for the Amazon calls for the creation of an inland sea, nearly as large as Western Europe! In Florida, the United States Army Corps of Engineers has built a series of small dams neatly across the northern boundary of the Everglades National Wildlife Refuge. This was done in order to irrigate land to be used for cattle grazing (notoriously, the least efficient use of land) and to appease the cattle-raisers' lobby. The result: the Everglades are drying up, wildlife is being destroyed, soil is becoming salinated, and parts of southern Florida are taking on the characteristics of a desert. To finish it off, there is a chance that a new jetport (with its high decibel levels and pollutants) may still be constructed at the southern edge of the Everglades.

We tend to overlook the fact that nearly all major disfigurements of the earth have been created by ourselves. The impoverished lands of Greece, Spain, and India, the man-made deserts of Australia and New Zealand, the treeless plains of China and Mongolia, and the man-made deserts of North Africa, the Mediterranean basin, and Chile, are all proof of the fact that *where there is a desert, man has been at work*. Ritchie Calder's *After the Seventh Day* documents this. It is instructive to compare maps of the United States, covering the period from 1596 to the present. Helpfully, the earliest maps, prepared by Spanish Catholic missionaries, are of the Southwest. The desert—which now covers parts of 9 states—hardly existed at all. But as trees were cut down indiscriminately, as water run-off increased, as an estimated 200 million buffalo were eradicated, as topsoil was washed away each spring, the familiar Dust Bowls of 1830 and 1930 were created, and the deserts kept growing. The only thing that has changed is the pace of change itself. It took Alexander the Great and other conquerors nearly 1,500 years to turn Arabia and Palestine ("land of milk and honey"), into a desert. A mere 300 years sufficed for the American desert. And American "know-how" has succeeded, through the use of defoliation, napalm, and the diversions of rivers and streams, in altering the ecological cycle of the southern part of Vietnam in five short years in such a manner as to ensure that part of the world turning into a permanent desert.

Of course, damage done by American know-how is visible not only in foreign countries and in our remote past. The other day I read in the papers that the city fathers of Butte, Montana, are devising plans for moving the entire city to a nearby valley, so that the Anaconda Company may enlarge their strip-mining operations. And even in space: more than 8 years ago a team of American military "scientists" exploded a charge of crystals in the Heaviside layer of the upper ionosphere. This was done—in spite of strenuous and vociferous objections by scientific bodies all over the world —just "to see what would happen." Inestimable genetic damage may have already been done to mankind, animals, and crops as a result of this experiment: *and we will never know, as there is obviously no control group available to us to measure the results, because all earth would be affected.*

It would be tedious to continue piling example upon examples, statistic upon statistic. For at a certain point, a feeling of deep lethargy sets in, and our reaction is: "What's the use?" or "What can one man possibly do?" If we respond in this manner, we are lost. For it is precisely by shifting problems from the trivial to the tragic plane, by forsaking a personal view for a cosmic one, that we rationalize and manage to shrug off our own personal responsibility. In *all* the above problems and more, all people are implicated, but not equally so. The designer's responsibility and implication is far greater. He is trained to analyze facts, problems, systems and to make what are at least inspired guesses regarding what may occur "if this goes on."

As of December, 1970, Los Angeles was the first place in which the total acreage used for roads and parking places *exceeded* the amount of space given over to human habitation. Obviously, the automobile is highly inefficient in many ways, and what is called for is a *designed* solution.

Recently, devices have been designed and marketed which cut down the exhaust fumes of a car. The installation of these exhaust filters has been made mandatory in some countries (Sweden) and in some states (California), and at a superficial glance, this seems an answer to the problem. But in fact it is not. The consumer is required to spend more money for installing one of these gadgets, and to continue spending more money since the gasoline consumption of the automobile rises sharply. Finally, the device itself is quite inefficient. Even this could be justified, if the automobile performed in satisfactory ways in all other areas, but it does nothing of the kind.

The answer here must inevitably lie in a complete re-thinking of *transportation as a system*, as well as a rethinking of each component part of that system. Some possible guidelines for the future already exist.

It is more than half a century now since a monorail rapid-transport system was first put into daily operation at Wuppertal in Germany. The system has proved itself to be fast and clean, and intrudes only minimally upon the physical and visual environment. Surely, monorail systems could help ease traffic congestion in many of our large city-smears. Furthermore, a technology that can send men to the moon has surely managed to discover devices better than the monorail during the last half century.

Nonetheless, we are told that the average individual in the Western world values his personal and individual transportation device, and that, especially in the United States, the family automobile has become surrounded by a whole cluster of ideas—relating to self-reliance, independence, and mobility—that once surrounded "Old Paint," in the days of the wild and woolly West. This folk mythology works only as long as we consider the automobile a sort of super-horse, and blind ourselves to its drawbacks. Once we consider the car as merely *one* link in a total transport system, alternative solutions can easily be found.

The average American today will drive his car (spatially the equivalent of 2 telephone booths or a large toilet) a distance of 60 feet around the corner just to mail a letter. On a second scale, he and his wife will drive one mile or so, once a week, to do their marketing. On a third level of complexity he may drive (quite alone in his huge steel coffin) a round trip of some 40 miles daily, to get to work. And on a fourth level, he may pile the entire family into the car (2 or 3 times a year) to visit Grandma some 300 miles away. A fifth level exists: he knows that at any time he has the ability to jump into his car, and by driving long hours, reach California from, say, New York in a mere 5 days. He rarely does this. He flies instead and rents a car in California.

Now let us analyze this as a system. Distances of more than 500 miles can be most efficiently transversed by airplanes. Distances of between 50 and 500 miles are more efficiently served by railroads, buses, monorail systems and other, newer methods, to be developed by design teams.

For distances of less than 50 miles, many devices now exist, some of which are not exploited sufficiently. New ones will be evolved by design teams. A partial listing in rising order of complexity seems in order. The simplest way to cover short distances still seems to be walking, and there is something ludicrous about millions of Americans driving a few feet to the mailbox, but solemnly "jogging" on a $276 aluminum treadmill in their bedrooms for 10 minutes every night. Roller skates may sound faintly ridiculous; they are nonetheless used in storage areas and to get around in factories in the space industry. Non-powered push scooters give excellent mobility to travelers arriving at Kastrup international airport outside Copenhagen.

An electrically powered aluminum scooter, weighing 18 pounds, foldable and, when folded, no larger than a shoebox, with a cruising range of 15 miles, was designed and tested by an industrial design student in Chicago several years ago. This device, which would give excellent mobility without pollution or congestion in downtown areas and on large college campuses, has never been built commercially. It would allow people to get from place to place on a platform measuring 9×15 inches (whereas the space occupied by a Cadillac "El Dorado" measures approximately $10 \times 19\frac{1}{2}$ feet); the saving in space is great. It is important to note that our industrial design student in Chicago worked alone for a period of 7 months and spent a total of $425 developing his electric, handbag-sized mini-scooter. Given the $3.4 billion which General Motors alone spends for corporate research and development each year, given the facilities and design talent available, we can readily see that even this excellent scooter is by no means the last word in personal transportation.

Bicycles are used for movement within the 50-mile radius we have established, both in Denmark and the Low Countries. Many of these fold, some can be carried easily. Powered bicycles with miniature gasoline engines exist; small electric drive-assist systems could be devised easily. Some of the vehicles designed by my students, for exercise and sport by both normal and paraplegic children, may point the way to new ways of transport. Mopeds, powered scooters, and motorcycles can be safely left out of this discussion in their present form as they are prime polluters. Finally, the automobile:

For reasons of prestige, "good taste," status, and sex appeal, as well as the easy profits guaranteed by built-in ob-

Electrivan: By 1968 there were more than 45,000 electrically propelled vehicles on the roads in Britain, more than anywhere else in the world. Without them and their extremely low running costs, Britons would no longer enjoy home milk delivery, garbage collection, ambulances, or street maintenance. The post office began using them some years back. Crompton Leyland Electricars have introduced this small and spunky van. It is 9 feet long, and has all the usual advantages of an electric: no clutch, gearbox, radiator or oiliness, which give low maintenance cost, plus a 20-foot turning circle and built-in charger. It can do 33 m.p.h. and will carry 500 pounds. It is guaranteed for 10,000 miles or one year. (Courtesy: The Council of Industrial Design, England.) So there seems little need for the controversy as to whether electric cars are feasible: thousands have been on the road for years!

solescence, few intelligent changes have been made in auto
motive design (aside from skin-deep styling and "conven
ience" factors) since 1895. Most of the configurationa
changes have gone in the direction of largeness and sleek
"zippy" looks. Nonetheless, a few design breakthroughs such
as the Bubble Simca, the two-passenger, in-line-seater
Messerschmidt, and even the Morris Mini-Minor and Mini-
Cooper, point the way to smaller vehicles that can, in the
last two cases, seat 4 adults and a child and also store an
incredible amount of luggage.

All that is needed is a new (probably electric) power
plant. Because batteries today are large in size, heavy, and
short-lived, such cars could recharge themselves from
sockets contained in parking meters and in their home
garages. Predictably, however, batteries will shrink in size
and weight and at the same time gain a longer life span as
industry discovers a need, and as the state-of-the-art pro-
gresses.

Some Utopian concepts such as moving sidewalks must be
rejected at this time because the power expended versus
the value gained is disastrously overbalanced in favor of the
former.

By joining three systems, all of them in existence right
now, we can find at least one viable alternative to the downtown clutter and traffic problem. If we combine (1) a fleet
of battery-driven miniature taxis similar to the Messerschmidt with (2) a transportation credit card and computer
billing of users at the end of each month and (3) a one-way
wrist-watch-sized radio, we have the beginnings of a rational
downtown transportation system. The user could summon
such a mini-taxi to his particular location with his radio
(thus eliminating the biggest argument against public transport: a long walk in the rain, and then a wait at the bus
stop). The mini-taxi could then take him to his specific location, again eliminating "approximate destinations." Payment
would be via credit card and billed monthly. Even with
thousands of these mini-taxis in downtown areas, more land
(now given over to garages, parking lots, and service stations) would be liberated. Exhaust fumes would be elimi-

nated. Larger parts of the streets could be given over to planting, parks, and walking space. At the end of the work day, users would be brought back to downtown monorail terminals, and returned to their home destinations.

Those romantic souls who would still prefer to "shift their own gears" and feel the soft purr of a high-powered sports car at their command, would find themselves in a position analogous to that of horseback riders today. At garages, located in a peripheral circle around larger cities, station wagons, trucks, or open sports cars could be rented for a few hours, or days, of country driving. This equipment could also be rented through the transport credit card system. However such vehicles could not be brought into built-up areas or cities.

(It is important to stress that the foregoing scenario is highly speculative and in no way pretends to present *the* answer to urban transportation problems. It merely attempts to show one of many possible solutions and, at the same time, deliberately shows how the designer and the design team are involved along every step of the way.)

The highest density of work population in Manhattan existed until recently at the junctures of Forty-second Street, and Madison, Park, and Lexington Avenues. Here some of New York's largest skyscrapers absorb a daily working population of more than half a million people, together with food, goods, and services for this area alone. Some of the largest shops exist within a one-mile radius. At Grand Central Station, 9 different levels of subways converge. The terminal building for 2 of New York City's 3 airports is only half a mile away. And Grand Central Station itself consists of 5 subterranean levels, and is the terminal point for nearly all railroads. The congestion and crowding are enormous. The answer of designers and planners to this problem was to erect a 46-story office building (equipped with a helicopter landing roof) directly over Grand Central Station! (This feeds an *extra* 120,000 people into the system each day.) While we are suggesting that designers make a positive commitment to today's problems, environmental or otherwise, the very least they might do in cases such as this would be to refrain completely from applying their talents and to refuse to participate in such insane and destructive acts.

Often the designer has controlled, or partially controlled, selection of materials and processes. For instance, the choice of aluminum as a better material for beer cans has been inaugurated by the merchandising staff of Alcoa. The fact remains that designers created the cans and the new "zip-openings" on them, which make them so attractive to the public. Industrial designers developed the cans, creative problem-solvers came up with the new opening method (and machines for production), and visual and graphic designers concerned themselves with brand identity, corporate identification, labels, trademarks, and selling the entire package to the public. What's wrong with that?

For one thing, the process wastes millions of tons of precious raw materials *that can never be replaced*. But more importantly, aluminum is a material that breaks down very slowly. For nearly a thousand years we will have to live with the beer cans thrown in the garbage today, or tossed casually out of an automobile last night. I have discussed Swedish experiments to develop a disposable beer bottle made of a bio-degradable plastic in Chapter Five.

Beer bottles and other canned goods are not the only offenders. Aluminum foils, while thinner, are every bit as resistant to rust, corrosion, and biological breakdown as cans. Used aluminum foil clutters up our dumps and acts as an effective shield against "breathing" by the top layer of the soil. This in turn directly affects rain absorption as well as the course of subterranean streams and natural water reservoirs. Soil temperature levels under dumps differ from those of adjacent areas by as much as 3° F. As well as creating minor climatic changes within small ecological systems, this also creates a tendency for garbage-shielded dumps to retain vital minerals, and prevent their absorption by the adjacent useful farmland. Here again imaginative designers should be able to suggest alternates to this system.

The introduction during the fifties of "aerosol" cans for most liquid and semi-liquid items under pressure has revolutionized merchandising of drugs, foods, home remedies, cosmetics, and many other items. Industry has embraced the aerosol concept eagerly: it makes it possible to sell a smaller quantity at grossly inflated prices. Almost without exception, aerosol cans are so constructed that even the small amount sold to the consumer cannot be totally used. Hence, more waste. Aerosol cans manage to foul up the landscape quite as thoroughly as others, but in addition, they are potential bombs, ready to send jagged pieces of metal shrapnel tearing into the flesh of any child incautious enough to experiment with the cans under heated conditions. Designers, both industrial and graphic, are much to blame in helping with the introduction of aerosol cans.

They are even more to blame for not creating better alternative solutions to the problem. The accordion squeeze-bottle designed for the Imco Container Corporation in 1955, Egmont Arens's toothpaste bottle for Bristol-Myers of 1957, and various European systems of merchandising mayonnaise, caviar, mustard, cheeses, and other foodstuffs in tubes, point the way to better approaches. As all of these are plastic, bio-degradable materials could be used, and the consumer would benefit additionally by getting a fair measure for a fair price.

The recent design of insulating wall sandwich panels used in buildings employs a combination of glass fibers, asbestos fibers, and a "chew" of other fibrous products. As little as one fiber, if accidentally breathed in, can cause death or serious illness. While the workers are well protected by face-masks, the surrounding space must also remain clean. This has led to the installation of giant blowers—blowers which frequently feed out into the street. During the last few years, several people have accidentally breathed in this material and died as a direct consequence. In Manhattan itself the same thing has happened during careless installation of such panels in buildings or apartments (cf.: the documentation by Berton Roueché). Once again, at least a sizeable part of the responsibility lies with the designer, in designing products less liable to be mishandled or act as pollutants.

These are just a few miscellaneous examples; a list of this sort could go on almost indefinitely.

If we turn to a man-made environment for living, the

story is at least equally grave. Frequently we are shown the cold, inhuman, and sterile aspect of the apartment buildings erected shortly after World War II along *Karl Marx Allee* in East Berlin. But the differences between it and similar speculative mass housing erected by insurance companies around Greater New York, or even the most enlightened "community planning" practiced in some of the Scandinavian countries, is only one of degree rather than kind. Human beings and family units have become "components" to be stored away like carbon copies in the gigantic file cases that are today's tenements. When the cry of "urban renewal" is raised, the results are frequently less humane than the situation that originally gave cause for redesign. Thus, in a recently "renewed" ghetto area located in the southeastern part of Chicago, a series of more than 30 apartment buildings (each holding more than 50 family units), is strung out in a single 4-mile-long chain, neatly placed between a 12-lane superhighway (which cuts the development off neatly from the rest of the city) and, on the other side, a series of large manufacturing plants (with their perennially belching smokestacks) and a large municipal dump. In spite of all the old ghetto's faults, it did have a "sense of community," and that has been destroyed completely.

The inhabitants have no park, green spaces, or even lone trees within walking distance. Each family is alienated from the rest; nights find them cowering in their cell-like apartments while the juvenile street gangs exchange gunfire down below. In just *one* of these buildings, more than

View of Frank Lloyd Wright's Cloverleaf Housing project that was to be built at Pittsfield, Massachusetts, in 1942. By Permission of the Frank Lloyd Wright Foundation. Copyright © 1969 by The Frank Lloyd Wright Foundation.

one case of rape or assault occurs daily, and between 3 and 4 cases of murder or attempted murder each week! The ghetto has been verticalized neatly and turned into a series of skyscrapers. Visually all the buildings are identical and look like a series of cement slabs into which a child has carved an insufficient number of tiny windows.

This area is also completely divorced from even the most basic shopping needs. A supermarket and a drugstore are located about 500 feet from the northernmost of the buildings, and public transport is lacking. This means that an elderly woman living, for instance, in one of the buildings at the south end of the development, has a 5-mile walk (round trip) in order to do her shopping. Thus a mother of small children is effectively removed from supervising her offspring for a period of nearly 3 hours when marketing. But the design of these barns for the storage of unskilled laborers and their families is not that different from similar developments for the well-to-do, or the rich.

It is a strange paradox in the design field (at least in the United States) that as our families have become larger in size, and as the furniture and furnishing designed for us take up more space, the size of our houses or apartments, as well as of individual rooms, has shrunk unaccountably. As a family achieve the financial means to leave these megablocks of flats, they are propagandized into buying a "home of their own." These homes again lack any and all individuality, and are strung out in a manner most convenient for the speculative builder, his machinery, and the plumbing

Cloverleaf Housing plan. By permission of The Frank Lloyd Wright Foundation. Copyright © 1969 by The Frank Lloyd Wright Foundation.

and street network which he installs, rather than for the needs of the people. If the prospective buyer is less than enchanted by 600 fake New England saltboxes, identical in looks and materials and placed cheek-by-jowl, usually his only choice consists of moving to another development in which another 600 equally phony French Provincial huts rub shoulders.

Even on low price levels, in 1969, the price of one of these homes is (in the United States) approximately that of 3 years of income of their prospective tenants. Naturally, these homes are purchased under deferred payment plans, mortgage settlements, and the like, usually covering a time period of 20 or 30 years. Compound interest usually adds 70–100 per cent or more to the price, and the family almost invariably moves out after 3 or 4 years. Should a tenant be foolish enough to attempt to individualize his own home through landscaping, planting, and other improvements, his neighbors will frequently heap scorn and abuse on him. Whether or not they do, his taxes will very probably be raised.

But all this is wide off the mark. What should be of concern to the designer is the relationship between the home and the way in which people live today. "Saleability" of mass housing is usually measured by how closely the houses conform to a late-nineteenth-century ideal of the trellised, rose-covered cottage. The fact that our style of life in the early seventies of this century (including such factors as: minimally four separate systems of telecommunication, "automobility," and increased leisure time, as well as the fragmentation of the basic family structure) is never taken into consideration. Nor the fact that the average American family moves approximately every 4.6 years. Most importantly, however, the basic human need to have some relation to green and growing things, as well as the option to do a bit of gardening or "farming" (for table crops such as tomatoes, lettuce, squash, and melons) is never taken into consideration. Nor is walking space, or play space for children, teen-agers, and adults.

Most designers (and not just in the areas of housing and

community planning) seem to have developed a set of blinders. This not only prevents them from generating new concepts, but (and this is crucial) *effectively keeps them from considering whether similar problems might not already have been solved intelligently somewhere else or at some other time.*

Say "Frank Lloyd Wright" to any shelter designer. He will immediately think of the Guggenheim Museum, Falling-water, the Imperial Hotel in Tokyo, and some of the earlier Prairie Houses. He may even think of a certain mannerist, neo-baroque interpenetration of space. But chances are that he will be totally unaware that Wright created an important "missing link" between individual homes and apartment dwellings.

In 1938, Frank Lloyd Wright designed the Sun Top Homes for Ardmore, Pennsylvania. Only one of the proposed four was actually built. It is really a cloverleaf-like interpenetration of 4 individual homes. Each one consists of a one and one-half story tall living room, and distributes a recreation room, bedrooms, kitchen, etc., over a 2-story area. Each individual quarter of the total 4-home construct is so defined that one is unaware of the other 3 units. The heating–air-conditioning–plumbing core is in the center of the complex. Nonetheless, each individual unit has its own air-conditioning, plumbing, and lighting facilities as well as its own kitchen garden, and a recreational garden, screened through trees and plantings from the other units and from the street. The entire building was extremely low in cost and actually built (as a first prototype) in 1941. In 1942 Mr. Wright further developed this concept for the Defense Housing Agency. One hundred of these cloverleaf homes (to house 400 families) were to be built at Pittsfield, Massachusetts.

The original Sun Top Homes prototype still stands at Ardmore, Pennsylvania, one third of a century later, mute testimony to the short-sightedness of the Federal Government.

The mix of heavy manufacturing offices, light industry, private homes, apartment-like shelters, clinics, day nurseries, schools and universities, sports arenas, recreational facilities, bicycle paths, access roads, forested areas, parking lots, shopping communities, and linkages to public transport and high-speed road networks which Frank Lloyd Wright designed in 1935 as Broadacre City, still marks a high point of humanistic planning. With local variations, Wright envisioned Broadacre City as eventually spanning the entire North American continent. Again, this is not to suggest that either Broadacre City or the Ardmore Development is the ideal answer.

According to Frank Lloyd Wright, *scale* was the greatest threat to social meaning. As early as the forties, he wrote: "*Little* forms, *little* homes for industry, *little* factories, *little* schools, a *little* university going to the people mostly by way of their interest . . . *little* laboratories. . . ." (Wright's italics)

Like Tapiola near Helsinki, Broadacre City and the Ardmore Development constitute partial solutions, but solutions more concerned with the quality of life and human dignity than the nearly 12 million rabbit hutches that have been built for human habitation since.

The whole concept of human scale has gone awry, not only with homes but in most other areas as well. One would expect a system motivated only by self-interest and private profit-making at least to spend some care in constructing its shopping places. This is not so. "*Strøget*," a "walking street" of shops in downtown Copenhagen, is constructed for leisurely strolling and impulse buying. Two segments of it, *Frederiksberggade* and *Mygade*, are together approximately 400 feet long and contain more than 180 shops.

In a contemporary American shopping center, this same distance of 400 feet will frequently separate the entrances of two stores: the supermarket and, say, the drugstore. The intervening space consists of empty windows, bereft of displays, monotonous and uninteresting. Usually neither landscaping nor wind-breaks are provided. Mercilessly, the hot sun beats down on the 4 acres of concrete in the summer; wind-whipped snow piles up in car-high drifts throughout the winter. Small wonder that, after finishing their shopping at the supermarket, people will walk back to their automobiles and drive to the drugstore. There is nothing in the environment that prompts going for a stroll; it has been designed for the moving car alone. Most shopping plazas in the United States consist of a thin line of stores arranged along three sides of a huge square, the center of which is a parking lot. The large open side fronts on a superhighway. This may make shopping "efficient" but it also makes it something less than satisfying.

This problem of scale is especially dramatic in our suburbs and exurbs, which have become vast dormitory towns with a multiplicity of problems.

Increasingly within the last decade, factories have moved away from large cities: cheap labor sources and large tax write-offs have induced them to move to so-called "Industrial Parks." (!) Foreseeably, more and more factories have begun congregating within each one of these so-called park sites. Around each of these clusters service industries, shops, and, eventually, tract housing has sprung up—without any plan, reason, or projection for future development. Transportation networks soon link these production centers to the old cities (crossing the suburbs and exurbs of yesteryear on their way). Soon a whole new subculture of minor assembly plants, repair shops, storage plants, etc., develops in the relatively large border area between city and suburbs, suburbs and satellite industrial centers. Without any rational plan, the city has just grown in area by a factor of 20 or 30 (due to the nature of this particular, omni-directional growth process).

Even if we are willing to accept psychological, social, and physical hazards in polluting our environment, there are other more immediate and weighty reasons for putting a stop to it. Recent information coming to us from weather satellites in space, direct observation in spaceships, and the statistics provided to us by meteorological observation stations quite clearly points to a major change. It seems that a large area of permanently polluted warm air will actually *attract* bad weather. In the American Midwest and East Coast especially, more and more major storms, snowfalls, blizzards, and tornadoes have, over the last ten years, hit large industrial cities. This phenomenon in turn (by increasing the number of target areas on the world's surface) in time may have lasting climatic effects. This is the curse of scale, when not attended to. As Julian Huxley remarks: "Simply magnify an object, without changing its shape, and without meaning to, you have changed *all* its properties."

Even the most basic study of systems design teaches us the obvious fact that a system is made up of its component parts, and that, as each part is changed, the system itself eventually will change too. By examining some of these systems we may be able to locate some of the factors contributing to the distortion. Hospitals and mental institutions are usually designed with greater care than other interior spaces. Architects, interior architects, and medical specialists routinely cooperate in the planning. In the floor plan, the rest-and-recuperation wing of a mental hospital at first may be well arranged for conversational grouping, relaxation, games, etc. But once the wing is put into operation, hospital personnel rearrange the seating at once. Chairs now are placed primly, neatly, and symmetrically. This has the virtue of bolstering the sense of security of the hospital personnel, cuts down the time needed to sweep and wash floors, and makes it far simpler for refreshment carts to be wheeled through the room. This furniture placement, however, creates fantastic barriers to interaction among patients, and in some cases may help to drive them into autistic or catatonic states. The placement of chairs on the four sides of each pillar, facing in four different directions, makes conversation extremely awkward for two people seated adjacently and completely shuts off conversation with anyone else.

This is no isolated example but rather something that happens in hospital wards all the time. It illustrates a cardinal error among designers: the failure to go back from time to time and see how the work has performed and been implemented. To my knowledge, hospital or mental patients have never been "client-group representatives" working with a design team. Similar observations can be documented regarding prison populations, the arrangement of living spaces

'or military personnel, university students in dormitories,
and other victimized groups held in captivity. Even in areas
where a profit-seeking system might be expected to pursue
its ends most efficiently, this is not often the case.

Edward T. Hall, in his studies of proximeters and human
spacing, has computed that the types and sizes of seating
units used in most contemporary airport terminals so
strongly violate Western concepts of spacing that fully one
third of them are empty at any given time. This holds true
even when the building is unusually crowded: many people
prefer standing or pacing to being brought into too close
proximity with strangers. Most Americans visiting Europe or
Latin America demonstrate signs of minor strain when asked
to share a restaurant table with strangers. And seldom has
Thorstein Veblen's theory of "conspicuous consumption"
been carried further than in the endless vestibules of motion
picture palaces, tenanted with gilt and scarlet chairs in
which no one ever sits, or in similarly appointed waiting
rooms to corporate offices (where oak and leather, steel and
glass have replaced the simpering charm of fake French
Empire).

Obviously, in each of these cases design decisions *have*
been made, but unfortunately, they were wrong. In each
case the designer has "worked up" a combination between
his personal aesthetic, the desires of his clients, and whatever has been considered "good taste" at the consumer level.
By working with a design team, checking conclusions
through our six-sided function complex, and working closely
with findings in the behavioral sciences, such mis-design
could be avoided.

Change has always been with us, but the dimensions
of this change are still not fully understood. As Alvin Toffler
said in *Future Shock,* "We are now living through the second
great divide in human history, comparable in size only with
that first break in historic continuity—the shift from barbarism to civilization. . . ." Much has happened within the
lifetimes of many of us that in sheer size is equalled only by
man's entire previous life on this globe. Half of *all* the

energy consumed by man during the past 2,000 years has been consumed within the last 100. The dividing line for many statistical series of materials (such as metals) seems to be about 1910. That is, man extracted about as much out of mines during the first 6 million years of his tenancy on this planet as during the last 60. The newspapers tell us that 25 per cent of all the people who ever lived are alive today; that 90 per cent of all the scientists and researchers who ever lived are living now; the amount of technical information doubles every 10 years; throughout the world about 100,000 journals are published in more than 60 languages, and this number doubles every 15 years. In the United States (a country notably exempt from the population explosion) the average density of people was one per square mile 200 years ago. In a circle 20 miles in diameter (assuming that 10 miles is the greatest distance a person can walk to work and back and still do a full day's work) this yields 314 persons inside the circle, with the rather obvious chance of human contact of 313 to one. Opportunities for interpersonal communication (the exchange of information and ideas) were formerly quite limited. Today, Chicago has a population density of 10,000 people per square mile, inside our 20-mile-diameter circle. Opportunities for human-to-human contact are more than 3 million to one.

Many of these changes are now setting a pace of their own, and are seemingly out of control. In Albert Romasco's *The Poverty of Abundance,* we find that: "In Ceylon, the introduction of DDT was largely responsible in less than 10 years for a 57 per cent decline in the death rate, a population increase of 83 per cent and a resultant decline in per capita income." It goes on to say, "There are nearly 300,000 babies born every day, two-thirds of them into families that are poor, hungry, ignorant, ill."

When we consider what a large population increase *does,* we find that it forces us into devising new ways of doing things because the old ones cannot work any more at all. A commission established in Tokyo to plan for the needs of a city of 20 million inhabitants within 10½ years soon found that *nothing in human experience, no technique adopted in*

*he past, is relevant to the kind of problems that arise when
ʼne considers the future's population densities.* Within less
han 15 years, several cities in India will have more than 36
million inhabitants each, and in the same time the United
States will add more than 100 million people to its popula-
ion. Within 35 years (present trends continuing) there will
ʼe close to 7 billion Africans, Asians, and Latin Americans,
constituting 86 per cent of the world's population. "If your
ʼulse beat is normal," says William Vogt, "it will not quite
ʼeep up with the increase in world population . . . Every
ʼime your pulse throbs, the population of the world has
added more than one human being." It has taken mankind
approximately 8 million years to reach a world population
ʼf 10 million people. It took another 12,000 years to reach
the one-billion mark; 75 more years to reach 2 billion; 37
years to reach 3 billion; and within less than 18 years we
shall have passed the 5-billion mark. Obviously, population
growth as a force for societal change has itself changed
from a *quantitative* force to a *qualitative* one.

We are beginning to understand that the main challenge
for our society no longer lies in the production of goods.
Rather, we have to make choices that deal with "how good?"
instead of "how much?" But the changes, and our awareness
of these changes, are becoming so highly accelerated that
trying to "make sense" of change itself will become our
basic industry. Moral, aesthetic, and ethical values will
evolve along with the choices to which they will be applied.
We may still consider religion, sex, morality, the family
structure, or medical research to be remote from technology
and design. But the margin is narrowing fast.

With all these changes, the designer (as part of the
multi-disciplinary, problem-solving team) can and must involve himself. He may *choose* to do so for vaguely humani-
tarian reasons (for maybe another 10 years or so in the
Western world). Regardless of this, he will be *forced* to do
so by the simple desire for survival within the not-too-distant
future. When you try to tell people in our Western society
that within a very short time, say 7 to 10 years, many of the

people in the world will die of hunger, they simply do no hear. They give a little nervous laugh; a little embarrassed they change the subject. But in Calcutta, Bombay, and New Delhi, thousands of bodies are already being removed by the sanitation squads each morning.

There was a time not very long ago, maybe in 1963, when as William Paddock put it, "The stork passed the plow." And now people are increasing faster than the means of feeding them. Less food is available per person in the world today than during the Depression some 30 years ago. Population is now increasing over food production at the rate of 2:1 per year.

Food production and the development of new food sources have been of no interest to the design profession at all. Yet designers *are* involved, like it or not, as human beings Raymond Ewell (editor of *Population Bulletin*) said a few years ago:

> If present trends continue, it seems likely that famine will reach serious proportions in India, Pakistan and China in the early 1970's, followed by Indonesia, Iran Turkey, Egypt and several other countries within a few years, and then followed by most of the other countries of Asia, Africa and Latin America by 1980. Such a famine will be of massive proportions, affecting hundreds of millions, possibly even billions, of persons. If this happens, as appears likely, it will be the most colossal catastrophe in history.

All the "concern" over the growth of the world's people only thinly veils violence and a sort of "escapism." It is no longer considered "nice" to be a racist. But the specific words many of us use when we talk about the people in developing countries, slums, ghettos, are bad. *Their* populations "explode," we say. *They* are a "population bomb." *They* "breed like flies." We talk about "uncontrolled fertility" and how we must "teach *them* to control population" and we talk (especially regarding Africa, Asia, and Latin America) about "breeding swarms." Such words reflect our thinking. And such thinking is our inheritance of racism, prejudice colonialism, white capitalist superiority and, when we begin to send "population control teams" to some country to "help," neo-colonialism.

Around 1800 there were an estimated 180 million inhabitants in Europe. The amount of people had increased to 450 million around 1900. But this fantastically increased population had a much higher living standard, ate better, dressed better, and lived longer than their own great-grandparents. The Malthusian Doctrine says: Food can *never* keep up with population growth. But this simple formula has just two factors: soil and population. Science, design, planning, research are completely left out. Malthus's theories may be applicable to animals (like laboratory rats), but the *one* function that is uniquely human, comprehensive anticipatory thinking and planning, changes his equation most drastically.

After all, only 90 years ago in the United States, a huge farm population (almost 75 per cent of the people) struggled desperately to keep a population of 85 million people from hunger. Today, *only 8 per cent of the population is still farming,* the population has surpassed the 200-million mark, and the biggest agricultural problem is what to do with megatons of food surplus each year! Agricultural machinery, irrigation, chemical fertilizers, scientific crop rotation, pest control, conservation, reforestation, selective breeding of stock animals—these are the fruits of science applied to Malthusian thinking, and they have destroyed his mechanistic concepts.

Naturally, families should not have more children than they can raise decently. *But birth control measures prove effective only after the living standards of the underprivileged have been raised.* That is the order: not the other way around. People begin to take interest in limiting the size of their families only *after* they are secure, have achieved human dignity and purpose, and are no longer beset by the anxiety and fears of hunger, poverty, ignorance, and disease. A large part of children born are no more than genetic insurance for people faced with the certainty of death for many of their children.

For hundreds of years we assumed what we were pleased to call "laziness" or languor, reduced energy, mental retardation, short life spans, and dullness to be racial characteristics in many underdeveloped countries. Today we know these are not races of lazy men; they are people chronically under-nourished, to the point where they are no longer energetic and hopeful. Malnutrition causes high infant death rates and often this occurs in families that have been made very large in the hopes of somehow compensating for this. But hunger and mental retardation go on, hand in hand.

> The brain grows more rapidly than the rest of the body, its cells proliferating so quickly that by the time a child is four, the circumference of its head is 90 per cent as large as it ever will be. . . . This proliferation is almost entirely dependent on protein synthesis, which cannot take place in the absence of the essential amino acids which must be derived from food. (*Bioscience*, April 1967)

Producing basic agricultural implements for under-developed areas of the world brings less profit to industry than producing glittering consumer gadgets for abundant societies. Designing agricultural systems and tools is not thought of as a "glory job" or "fun" by most designers: how much more rewarding to "scale down" a 1931 Mercedes SS for fiberglass production, than to improve a plow for Pakistan!

The most significant gains in farming can be made through design and systems analysis. In order to support this thesis, I shall quote extensively from the "Famine 1975?" issue of the *Kaiser Aluminum News** (from which most of the material cited above is derived). The headings of the pieces are: "Land," "Water," "Fertilizer," "Pesticides," "Preservation & Processing," "Livestock," "Mechanization," "Transport," "Marketing," and "Education." Each of these pieces is an input that fits and reinforces those around it.

> None can be solved on an isolated basis. And no single sub-problem offers the final solution to world hunger, anymore than a single radio can be called the answer to global communication. The world food crisis is a systems problem in which each part is amenable to solution, given the impetus of a sufficient political, social and economic commitment . . .

* "The World Food Crisis," *Kaiser Aluminum News*, Vol. 26, No. 1 (April 1968).

In many parts of Asia where water is critical, it would take 100 years to add one more cultivated acre per person. Meanwhile *the population will have increased 16 times.* If increase of cultivated land is to match population gains, then annual per capita expenditures in both water and land development must accelerate to more than 4 times the present level. Significantly improving crop yields is the only other alternative. But such double or triple cropping requires greater use of fertilizer, improved irrigation, and better pest control. The cost of bringing new land into cultivation varies widely, from $973 per acre in Kenya to $612 in the United States and $32 at one pilot project in Guatemala. Assuming an average cost of $375 per acre world-wide, the 4 billion arable acres remaining in the tropics alone would require an investment of $1.5 trillion.

The land study reaches the following conclusion: ". . . since we don't yet have an accurate knowledge of what resources are at our disposal, a worldwide inventory of soil and water, as well as capital, manpower, and technology available, is a prior need."

In 1963 Bucky Fuller began a World Resources Inventory at Southern Illinois University, in Carbondale. This design study group has published 6 of some 10 planned reports. However, only a skeleton staff of 5 people remains to complete the work because designers, students, and design schools have shrugged off the entire matter as "dull," "uninteresting," and "unimportant."

If we study water, we find that less than 11 per cent of the world's cultivated land is irrigated. Part of the problem is lining ditches to control seepage, recycling water, and waste removal. In an earlier chapter I have already discussed the bad circulation system of the African continent and tried to point out a few ways in which simple, hand-operated "pipe-making" tools might be used to help in irrigation and pollution at the village level.

Ground water estimated to be 3,000 times greater in quantity than the contents of all the world's rivers lies within half a drilling mile beneath the earth's surface. The Sahara Desert, for instance, contains 100 billion acre-feet of water in huge sub-surface aquifers; enough water to irri-

gate millions of acres for at least 4 centuries. To develop
tapping, drilling, and distribution methods is a new design
challenge. The desalination of ocean water is a process now
in use in Israel. Rational design should make it possible to
reduce costs.

Fertilization and pesticides and their influence on the
environment have been discussed elsewhere. But it is in the
area of preservation and processing of foods that designers
could make a major contribution.

> Food losses after harvest run as high as 80% in the diet
> deficient countries, due largely to poor storage and proc-
> essing. Micro-organisms, insects and rodents are the
> main cause of food loss after harvest. Rats consume 16
> times more food than humans per body weight; in India
> rats eat 30% of stored grains; in some countries as
> much as 60%. One-third of all harvested cereals in
> Africa is lost to rodents. Because of poor and out-dated
> equipment, lack of refrigeration and inefficient trans-
> port, 50% of marketed fruits and vegetables are lost in

A modular cooling unit for perishable foods, to be used in underdeve
countries. The unit is hand-cranked and can be made for under $6
signed for UNESCO by James Hennessey and Victor Papanek.

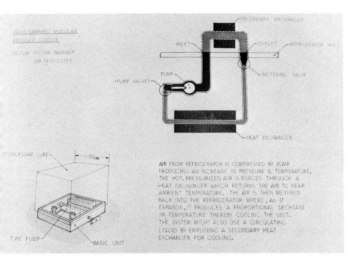

WIND CHARGED MODULAR
PRODUCE COOLER

DESIGN: VICTOR PAPANEK
JIM HENNESSEY

SECONDARY EXCHANGER

INLET   OUTLET   REFRIGERATOR WALL

PUMP

PUMP VALVES   METERING VALVE

HEAT EXCHANGER

STYROFOAM CUBE   50 cm

TIRE PUMP   BASIC UNIT

AIR FROM REFRIGERATOR IS COMPRESSED BY PUMP
PRODUCING AN INCREASE IN PRESSURE & TEMPERATURE.
THE HOT, PRESSURIZED AIR IS FORCED THROUGH A
HEAT EXCHANGER WHICH RETURNS THE AIR TO NEAR
AMBIENT TEMPERATURE. THE AIR IS THEN METERED
BACK INTO THE REFRIGERATOR WHERE, AS IT
EXPANDS, IT PRODUCES A PROPORTIONAL DECREASE
IN TEMPERATURE THEREBY COOLING THE UNIT.
THE SYSTEM MIGHT ALSO USE A CIRCULATING
LIQUID BY EMPLOYING A SECONDARY HEAT
EXCHANGER FOR COOLING.

the hungry nations, where most perishables must be
eaten within 24 hours of harvest.

There is little under "livestock" that relates directly to the
work of designers. But at present, livestock is a highly inefficient and costly way of obtaining proteins. Here the development and production of "single-cell proteins" will provide the breakthroughs. The production and equipment for
the laboratory farming of protein bacteria is well within the
design team's area of concern. Benefits are dramatic:

> The full potential of protein bacteria is easier to grasp
> if it is compared with a properly-fed 1,000 pound steer.
> The steer stores up just one pound of protein a day. In
> the same 24 hours, a half ton of selected micro-or-
> ganisms, feeding on oil, increase in size and weight by
> five times, and half this gain is useful protein. In other
> words, *while the steer is making one pound of protein,*
> *the bacteria-in-oil produces 2,500 pounds of protein.*
> (Professor Alfred Champagnat)

The present cost of this protein, because of an insufficient
state-of-the-art and badly designed equipment, is about 20¢
per pound. As to the taste:

The line of meatlike but meatless foods includes ham
sausage, frankfurters, fried chicken, steaks, meatloaf
chipped beef and luncheon loaves. They have no bones
skin or excess fat . . . surprisingly, most of these rated
high both in taste and appearance. (*Successful Farm
ing,* October 1967)

Mechanization is the next one of our puzzle pieces. Studies
have shown that to get high crop yields, mechanical energy
of about .5 horsepower is needed for every 2½ acres of
land. In the United States and in Europe, the energy level
is more than twice that. But in the hungry nations it is very
low: less than .3 horsepower in Latin America, under .2 in
Asia, and only .05 in Africa. The Kaiser report concludes:
". . . the most difficult task would be recruiting and train
ing the 10,000 designers needed. . . ." In this connection it
is worth noting that, of the 692 "professional" members of
the Industrial Designers Society of America (the *only* pro
fessional design group), only 18 are involved in the design
of farm machinery. Twelve of these are producing mini-
tractors and ride-em-yourself lawn mowers with which the
rich manicure their lawns. Only the remaining 6 are involved with prime producers of farm machinery. And even
their involvement consists largely of surface styling, trademark design, sexy promotion gimmicks, and more comfortable tractor seats.

Advertisement for a Mini-
tractor, by the Plastics Division of Kodak.

Transport seems least efficient where it is most needed. In India alone more than half a million villages are more than 5 miles from a road, and many of these roads are not passable in bad weather. One alternative to the costly development of the road networks and vehicles would be the development and wide distribution of ground effect vehicles that would make roads unnecessary (but would not be powered by internal combustion engines).

Note: the only ground effect vehicles now in operation are either designed to transport tourists and their automobiles across the English Channel, or else are being used by the United States Army to napalm people and villages in the Mekong Delta.

It is of additional interest to compare the genesis, development, and marketing of a relatively new consumer product. Beginning in the mid-sixties, firms in the United States, Sweden, West Germany, and a few other countries began introducing a series of gasoline-powered sleds on the market. These "snowmobiles" are largely bought and used in winter sports areas. They sell almost exclusively to wealthy and jaded young "athletes" whom these zippy little gadgets enable to traverse snow-covered terrain in "easy-chair comfort" with mechanical heaters helping them to "rough it." These self-propelled power scooters currently sell for approximately $995. Snowmobile clubs and rallies have been organized to help with the sale of these vehicles, and by now an attractive range of extra accessories (built-in tape decks, two-way radios, trailers, etc.) are available to the "discerning consumer." However, a new need has made itself manifest.

Rural populations in Canada and Finland, Lapp populations of Norway and Sweden, and the people living in the polar regions of the Soviet Union have found that vehicles of this type are useful in hunting, fishing, herding reindeer, and for emergency transportation needs. Surely here is an area in which industrial designers could work with their traditional values of cost reduction and mass production. Such a vehicle, selling for around $100 or less, would provide a new working tool for a large number of people now living under marginal conditions. Instead, snowmobiles are

becoming more complex, more loaded with consumer value
and still higher in cost.

A student design team of mine is now attempting t
develop and build a first prototype of a low-cost, battery
driven snowmobile, specially designed for the needs of th
Eskimos, Northwest Coast Indians, and Lapps. Before start
ing on this project, we discussed it with relevant sectors c
American industry who dismissed the entire approach a
impractical, silly, and "not needed." We shall also attemp
to make our scooter nearly noiseless and free from pollutin
agents.

"Education" forms the concluding part of the Kaiser Re
port. A country's agricultural development is related directl
to a rising level in general education among its people. I
developing countries there are ¾ of a billion illiterat
people. The fact that there are 20 million *more* illiterates i
these countries today than there were only 5 years ag
shows the inadequacy of educational efforts. My own low
cost transistor radio, pictured and described elsewhere i
this book, serves as a good example of how one smal
somewhat "gadgety" device may become a link in a tota
educational system and thereby transcend the narrow rol
is was originally designed for.

It is criminal that at present no area of design fo
agriculture forms even a small part of any curriculum of de
sign taught at even one school anywhere! Instead of ad
dressing themselves to such environmental needs, the in
dustrial design schools are making a concerted effort t
teach design for settings far more exotic.

During the spring of 1969, six leading American design
schools were involved in a competition and exhibit to design
housing and working environments for the ocean floor. The
heavy publicity surrounding these endeavors was nearl
drowned out by another program which concerned itsel
with the design of an entertainment center to be erected o
the moon. There is little doubt that soon men will have t
harvest the protein-rich fields that are the world's oceans
Nor will it be long before we drill for minerals and oil o

ocean floors and farm the fish and algae of the seas. And certainly before the end of this century men will look to the stars while living in semi-permanent domed settlements on the moon. But the necessities of today cannot be neglected for the expediencies of some dubious tomorrow. Design competitions such as the two mentioned above are usually assigned because they are more glamorous, "glory jobs," more fun than coming to grips with real problems. It is also in the interest of the Establishment to provide science-fiction routes of escape for the young, lest they became aware of the harshness of what is real.

Designs will be needed when man establishes himself on our ocean floors and on planets circling distant suns. But man's leap to the stars and his life beneath the seas is heavily conditioned by the environment we create here and now. There is something wrong when young people are less familiar with life on a southern Appalachian farm than with the construction of a gambling casino on Mars. They are taught a lie when they find themselves more familiar with atmospheric pressures in the Mindanao Deep than with atmospheric pollution over Detroit.

11

THE NEON

BLACKBOARD:

The Education of Designers

and the Construction of

Integrated Design Teams

Telling lies to the young is wrong.
Proving to them that lies are true is wrong.
The young know what you mean. The young are people.
Tell them the difficulties can't be counted,
And let them see not only what will be
But see with clarity these present times.
YEVGENY YEVTUSHENKO

EDUCATION FOR DESIGNERS (like nearly all educa-
tion) is based on the learning of skills and the acquisition of
a philosophy. It is unfortunate that in our design schools
both of these are wrong. The skills we teach are too often
related to processes and working methods of an age just
coming to a close. The philosophy is an equal mixture of
the kind of self-expressive bohemian individualism best expressed in *la vie bohème* and a profit-oriented, brutal com-
mercialism. Moreover, at best the "method" of transmitting
all this is almost half a century old.

In 1929 the Albert Langen Verlag of Munich published
the book *Von Material zu Architektur* by László Moholy-

Nagy as Volume 14 for the Bauhaus. Moholy-Nagy attempted to find new ways of involving young people with the
interface between technology and design, design and the
crafts, design and art. Possibly the most important idea was
to have students experiment directly with tools, machines,
and materials. Technological development did not end with
the electric band-saw sheltered in some Bauhaus basement
in 1919. To base education on this kind of development is to
ignore data-processing, computer technology, remote handling mechanisms, jet airplanes, space research: in fact all
that science and research have developed for mankind in
this, their most productive half century.

When Moholy-Nagy started the "New Bauhaus" (later the
Institute of Design) in Chicago, the book was republished
(by Norton in 1938) under the title *The New Vision.* An
expanded and lavishly illustrated re-hash of all this was
brought out shortly after Moholy-Nagy's death under the
title *Vision in Motion* in 1947. And now, nearly a quarter of
a century later, this 1947 re-hash of a 1938 translation of a
1929 book describing design experiments carried on in 1919
still forms the basic design curriculum at nearly all American and European schools. Happily, experiment turned into
tradition marches into the second half of the century. Can
we wonder that students are bored? Can we wonder that our
young people no longer consider the university and its
courses relevant to living? Surely a student entering a design school or university in September of, say, 1971, must be
educated to operate effectively in a professional world *starting* in 1976, and foreseeably he will reach the height of his
professional competence in 1995, or the year 2000.

Learning must be an ecstatic experience, as George B.
Leonard maintains in *Education and Ecstasy.* At best, for
instance, learning to drive a car is ecstatic (as any sixteen-
year-old will tell you). To drive an automobile demands a
fantastic combination of motor coordination, physiological,
and psychological skills. Watch the thousands of people
driving along the Los Angeles Freeway at 5:00 P.M. any
afternoon. People controlling 2 tons of steel and machinery,
hurtling along at better than 60 miles per hour, with the

distances between cars to be measured in inches. It is an impressive performance. It is a *learned* skill. And it just possibly may be the most highly structured non-instinctual activity these drivers engage in in their entire lives. They drive superbly well; the clue to their performance lies in the original method of learning how to drive. For to learn is to change. Education is a process in which the environment changes the learner, and the learner changes the environment. In other words, both are *interactive*. Both the beginning driver and his car, as well as road system, other cars, and his teacher, are locked into a self-regenerating system in which each slight perfection of every slight skill is immediately rewarded or positively reinforced. To return to George B. Leonard (on page 39 in *Education and Ecstasy*):

> No environment can strongly affect a person unless it is strongly interactive. To be interactive, the environment must be responsive, that is, must provide relevant feedback to the learner. For the feedback to be relevant, it must meet the learner *where he is,* then program (that is, change in appropriate steps at appropriate times) as he changes. The learner changes (that is, is educated) through his responses to the environment.

Unfortunately, education has been made into a method of preserving the status quo, a way of teaching and maintaining the moral attitudes, smug life-styles, and other sacrosanct values held by the old, and dispensing whatever is currently accepted as "Truth."

Digressing for a moment, we find that the example of learning to drive an automobile, cited above, is really just a scale model of how mankind as a whole has learned to live. For through millions of years man was a hunter, a fisherman, a sailor-navigator. As a hunter, he roamed earth as a member of a small hunting party—a cross-disciplinary team, in a way. He evolved early (but elegantly functional) tools: evidence from Choukoutien in China shows that Peking Man (*Pithecanthropus pekinensis*) fashioned stone tools long before *Homo sapiens* appeared on earth, and used fire as well.

Man as a hunter-fisherman-sailor was a non-specialist or a generalist, whose brain furnished him with that social understanding and control of casual impulses needed in a hunting group or society. We are told that even language evolved in answer to group need in the hunting party.

As hunter, man was highly successful. Equipped with spear-thrower, slingshot and bow, with knives superbly crafted of obsidian, horn, or bone, he spread from Siberia to Spain and from the ice cliffs of Afghanistan to Mesopotamia. And adventuresome early hunters followed bison and mammoth across the frozen Bering Strait into North America, where they settled the Great Plains nearly 15,000 years ago. They were *Homo sapiens*, and they were hunters. *Farmers would never have survived.* Even the art works of the Upper Paleolithic are evidence of a fairly leisurely existence and, in Europe at least, life may have been quite pleasant for these hunters.

I am not suggesting the hunter as a "noble savage" *à la Rousseau*. Compared with his farmer-descendant of the Neolithic, he may have been a rough, nearly savage fellow. Yet as we study Paleolithic archeology or read about and live with the disappearing tribes still essentially Paleolithic today (the Bushmen of the Kalahari, the Australian aborigine or the Eskimo), we see much that is ingenious and admirable.

To quote from Nigel Calder's *The Environment Game:*

> How do you deal with an angry bull elephant, when all you have is a sharpened stone? You nip aside, slip in behind, and cut the tendons of his heel. What can you do to lure a giraffe, the most timid of large animals? You play on its curiosity for bright objects by flashing a polished stone in its direction. The Bushmen, according to Laurens Van der Post, would use lions as hunting "dogs," letting them kill game and eat a little, before driving them off with fire. Franz Boas tells how Eskimo approach deer, two men together, one stooping behind like the back end of a pantomime horse, the other carrying his bow on his shoulders to resemble antlers and grunting like a deer. The despised Australian aborigine can "travel light" with only a few wooden and stone im-

plements and, by his knowledge of nature, survive indefinitely in the Great Sandy Desert. If we once let these echoes of our pre-history penetrate our sophisticated heads, they strike in us chords of excitement, if not of envy.

Traditionally we are taught to see farming as the prerequisite of civilization. An elaborate social life, we have been told, could not develop until man was freed from the daily chore of fishing or hunting. Lately however, this theory is being challenged by the view that early civilized settlements were based on highly organized food gathering rather than cultivation. The highly structured societies of American Indians and the salmon-eaters of British Columbia were so well supplied with food that large settlements developed.

To return to Nigel Calder in *The Environment Game:*

> Man's chief physical disadvantage as a hunter must have been the encumbrance of his family. The human infant is uniquely helpless and slow to mature. Accordingly, a fairly settled, well-defended domestic life was necessary from the outset. Women at home minding the children while the men were out hunting were well placed to develop arts like cooking, clothes-making, and pottery, to experiment with new foods, and to discover in their "gardens" the elementary principles of plant reproduction. Jacquetta Hawkes has remarked, "It is tempting to be convinced that the earliest Neolithic societies gave woman the highest status she has ever known." (*Prehistory,* UNESCO History of Mankind)

In truth it was agriculture that turned man towards the fateful downward slope of specialization. Mankind, heretofore dynamically moving through the environment as a member of a non-specialized, cross-disciplinary hunting party, now settled down to patient, millennia-long cultivation of the soil. Instead of learning through interaction with the environment, he substituted eons of boredom and elevated tradition to wisdom; hence to be conservative was a virtue. With human settlements located in prime agricultural areas, natural disasters became major destructors of the social pattern. Zealous and vengeful gods had to be appeased through priestly classes, sacrifices, and rituals. Man no longer stood and fought his surround alone, moving freely across the globe. Instead, territory became precious and war an extension of statecraft.

As Buckminster Fuller has said, every living creature is more specialized than man. Most birds can fly beautifully, but find it almost impossible to walk. Pigeons can do a little better at walking than most other birds, and the robin can hop. But most birds can't begin to walk at all. Fish swim beautifully, and get along in their medium, but they can't walk and (usually) can't get out on the land. These are all highly specialized forms of life. What is absolutely unique about man is his ability to apprehend, comprehend, and employ information, and to undertake unprecedented tasks.

For millions of years man's "little red schoolhouse" was earth itself. Mankind was taught to react and to behave by the environment, disasters, and predators. But now we have replaced our "natural enemies" with *educators,* and we try to learn from them. To brutally twist man away from his natural heritage of non-specialization in this way can only have brutal results. It is in the area of driving men into ever-narrowing fields of specialization that the schools and universities have made their greatest mistakes. Today's "revolution on the campus" is the students' intuitive reaction.

Modern technology (computers, automation, mass production, mass communication, high-speed travel, etc.) is beginning to give mankind a chance to return to the interactive learning experience, the sensory awakening of the early hunter. Hydroponic farming, "fish-herding," protein manufacture, and skyscraper farms will help also. Education can once again become relevant to a society of *generalists,* in other words, designer-planners. We have established in the first chapter that designers (especially) must operate on a non-specialized basis; it is little wonder, then, that the intuitive student revolution against "status quo education" seems often to happen first in our schools of design. For the designer shapes the environments in which we all live, the tools which we all use. And from the unpalatable manifestations of bad design in our society, the design student cannot remain aloof for long.

The main trouble with design schools seems to be that they teach too much design and not enough about the social and political environment in which design takes place. It is impossible to teach anything *in vacuo,* least of all in a system as deeply involved with man's basic needs as we have seen design to be.

To this dichotomy between the real world and the world of the school, there have been, understandably, many different answers.

It's a wondrous thing that schools of industrial design come in all shapes, sizes, and flavors, and that, in the United States alone, there are 62 of them. Most of these educational mini-sections are small enough that the program is determined by the departmental chairman. Unavoidably a "Cult of Personality" has grown up in each of these schools. And while this has not always been good, it at least has encouraged diversity.

Securely sheltered under Chicago's umbrella of air pollution, there flourishes a school that specializes more and more in establishing better communication between the student-designer and a computer. How far this experiment may lead to better communication between designer and user still remains to be seen. The sexy and slick presentation-drawings emanating from several California schools may prove valuable to a re-make of *2001: A Space Odyssey* and after all, "What's good for General Motors is good for the country!" And then there are the many other schools where dedicated young men and women build balsa-wood and plaster models of such important tools of everyday life as transistorized back-scratchers, electric hairbrushes, and dual-purpose electronic climate boxes (one half of which humidifies bread, while the other half de-humidifies cookies and crackers).

"Artsy-craftsy" schools still predominate throughout northern Europe, the Scandinavian countries, and much of the

United States. Here students still are taught to hand-hew little fetish symbols out of silver, teak, or hopsacking, or to create the *one* perfect ceramic teapot (when it is plain that entirely too many different types of teapots are in existence already).

That medical instruments should be kept sterile goes without saying. But that medical instruments must also *look* sterile, and that in fact appliances, clocks, toasters, and kitchen scales should also reflect all the romantic values of a surgical sterilizer—this is unhappily not restricted to just a few German schools.

A mish-mash of such courses as "Design Methodology," "Motivational Market Consumer Testing," and "Psychological Eidetic Identity" still clutter up the program of many schools. And overshadowing all of this is the sacred concept that things must be designed to appeal to temporary fads, fashions, and the acquisitive instinct; that they must be designed to wear out; and that they must make as much profit as possible for the company and its stockholders. Naturally, all these good young people may be revolted, and hence revolt when asked to fake up yet another bathroom scale, alarm clock, or fire extinguisher.

After all, students could argue that, inasmuch as we have succeeded (nationally) in the murder, rape, torture, pillage, and genocide of some 60 million Indians; and as the nations of the world have succeeded (internationally) in murdering, napalming, atomizing, and maiming some 150 million people *during the last 54 years alone;* and as another 600 million men, women, and children (one sixth of humanity) are starving to death, or dying of easily curable diseases within this decade; that somewhere *we* (the designers of our environments, our tools, and our products) have been missing the boat.

Yet the designers' "responsible" answers to such crises (last year alone) have consisted in such trivia as a $3,000, 14-carat gold toilet seat, appetizingly advertised in the newspapers of La Jolla, California, a wall-to-wall bathroom rug made of monkey fur and selling for $12,800, and finally, for the economy-minded, that life-sized, inflatable plastic woman at $9.95, or $16.95 for the deluxe model (see the advertisement in *Argosy*, February 1969, page 93; as well as Chapter Six of this volume).

In a foundation-endowed society (with the "Universal Credit Card" just months away), the foregoing smacks of heresy. Why, aren't our think-tanks thinking about the unthinkable? Don't we produce the finest artificial grass? Aren't our fiberglass rocks (hollow, for easier moving) the envy of the Free World?

Well, what *is* the position of design in the West today? We know that the twin concepts of "designed aesthetics" and "designed obsolescence" are heavily interrelated, and this connection becomes very apparent both in basic research and in the manufacturing process. Objects are designed, made, and purchased in a variety of styles. A French Provincial TV set, a Baroque refrigerator, or an "Early American" skyscraper strikes few consumers, or for that matter even designers, as anachronistic and silly. Even within the narrow band of "modern" or "contemporary," many different stylistic approaches exist and are accepted by the public. The clear direction of the past has changed into erratic and random fragmentation.

One cause of this fragmentation lies in our economic processes. Consumer goods of every sort, including houses, apartment buildings, civic centers, and motels, must seem continuously new. For we buy or rent only that which is changed and, moreover, *looks* changed. Industry, hand-in-hand with advertising and marketing, teaches us to look for and recognize these superficial changes, to expect them, and, ultimately, to demand them. Real changes—basic changes—mean retooling or rebuilding; in our present system the costs of this are prohibitively high. But to repaint and/or rearrange surfaces (interior or exterior) is just as exciting to the propagandized lay public, and can be done more cheaply.

Thus, the vital working parts of a mechanism (the guts of a toaster, for instance) will remain unchanged for years while surface finish, exterior embellishments, control me-

chanisms, and skin color and texture undergo yearly muta-
tions. This will hold true even if the working part is far from
perfect or in fact has major weaknesses or faults (as in the
case of automobiles, motor boats, air-conditioners, refrigerators, or washing machines). Automation also tends to make
periodic re-evaluation of the real design problems prohibitively expensive. The regional planner has become a landscape designer, the architect a decorator, and the designer a
stylist or cosmetician. Mechanism and structure are relegated
to the appropriate engineer and the product lacks all unity
or wholeness of purpose.

Accidentally, even the lowly stylist may strike some common associational or telesic chord that makes the consumer
wish to hold on to the product, rather than trade it in for
the latest version. (Recent examples of this are the 1961
Mustang and the 1954 Porsche.) To break down even this
accidental unwillingness on the part of the consumer to
throw away things, we have evolved materials that age
badly. Throughout most of human history, materials, being
organic, have aged gracefully. Thatched roofs, wooden
furniture, copper kettles, leather aprons, ceramic bowls—
all these and more would acquire small nicks, scratches, and
dents, gently discolor and acquire a thin patina as part of the
natural process of oxidation. Ultimately, they would disintegrate into their organic components. Today we are taught
that ageing (be it of products or individuals) is subtly
wrong. We wear, use, enjoy things as long as they look as if
they had just been bought. But once, under sunlight, the
plastic bucket deforms (however slightly), once the fake
walnut table-top melts under a cigarette, the anodizing on
an aluminum tumbler slips, we are taught to throw the of-
fending object out.

This divorce between the working mechanism (which, because of tool and die-making costs, remains unchanged) and
the more and more evanescent skin surface has led to
further specialization and to an aesthetic based on outward
appearance only. The "skin" designers (Detroit's stylists)
disdainfully avoid the "guts" designers (engineers and research people); form and function are split. But neither a

creature nor a product can survive for long when its skin and guts are separate. Finally, basic design research seems unsound because of the huge effort needed to keep up with a rapid technology. The knowledge and care which the product itself demands are diverted.

A more durable kind of design thinking sees the product (or tool, or transportation device, or building, or city) as a linear link between man and his environment. In reality we must think of man, his means, his environment, and his ways of thinking about, planning for, and manipulating himself and his surround as a non-linear, simultaneous, integrated, comprehensive whole.

This approach is *integrated design*. It deals with the *specialized* extensions of man that make it possible for him to remain a *generalist*. Such means and extensions already exist, but if we wish to relate the human environment to the psychophysical wholeness of the human being, we have to develop new, modified, and growing extensions and means on several new planes. Our goal would be to replan and redesign both function and structure of all the tools, products, shelters, and settlements of man into an integrated living environment, an environment capable of growth, change, mutation, adaptation, regeneration, in response to man's needs.

All man's functions—breathing, balancing, walking, perceiving, consuming, symbol-making, society-generating—are completely interrelated and interdependent.

Integrated design will concern itself, for the first time since the Late Paleolithic, with *unity*. This must include regional and city planning, architecture (both interior and exterior), industrial design (including systems analysis, transportation and bionic research), product design (including clothing), packaging, and all the graphic and film-making skills that can be generally subsumed under the catch-all phrase of visual design. Dividing lines exist between these areas at present, but the lunacy of these divisions is apparent even on the most basic level. To use one example: What is architecture? Assuredly it is more than the skill of building arches. Considering today's mix of civil engineering, speculative building, contracting, interior decoration, federally subsidized mass housing, landscaping, regional planning, rural and urban sociology, sculpture, and industrial design, can architecture even be said to exist as a separate discipline at all?

Certainly the "formal" grammar of building types can be said to have been enlarged greatly during the last 50 years. Nervi and Catalano have given us new ways of dealing with pre-stressed and reinforced concrete shells. Jim Fitzgibbon and Bucky Fuller have given us Synergetics, Geodesics, the Dymaxion House, and the dome. Bruce Goff and Herb Greene in the United States have developed a whole new concept of indigenous building. Bill Katavolos and I have both played with theoretical methods of *literally growing buildings organically*. But none of this has really enriched the field of architecture—whatever that may be. For we are able to compute almost exactly the wind forces operating against Herb Greene's Prairie Houses and define weight-to-cost ratio in a Bucky Fuller dome, but we have done little or no basic research assessing in what kind of a structure the human organism lives, works, interacts optimally. There is not enough knowledge regarding some of the most fundamental aspects of architecture.

In other ways, architecture can hardly still be considered an area of its own (it lacks definition), and, finally, it overlaps with dozens of different fields. In view of all this, what is architecture? Could this be the reason why so many architects have moved towards city planning and industrial design during the last decade? And during that same time, industrial designers have concerned themselves increasingly with the development of prefabricated houses and building components. Interior designers have developed furniture, tools, and are currently caught up in the fad of "Super Graphics" while visual designers develop products and make films.

There is a sort of Brownian motion going on throughout all the separate areas of design, and I believe this to be an intuitive response to dynamically changing times, similar to the intuitive dissatisfaction and unrest of students. Surely it would be more rational to say that within the field of integrated design, many different levels of complexity exist. These might concern themselves with the relationship of human and structural factors in a material (or a set of materials) that provides shelter or with the interaction between a transportation device, a road network, and the landscape.

If we speak of integrated design, of design-as-a-whole, of unity, we need designers able to deal with the design process comprehensively. Lamentably, students so equipped are not yet turned out by any school. For their education will be less specialized and take onto itself many new disciplines now thought of as only distantly related to design, if related at all.

Integrated design is not a set of skills, techniques, or mechanical processes, but should be thought of as a series of biological functions occurring simultaneously rather than in a linear sequence. These simultaneous "events" can be thought of as initial fertilization, developmental growth, production (or mimesis), and evaluation, the latter leading to re-initiation or regeneration or both, thus forming a closed feedback loop.

Integrated design (a general unified design system) demands that, through careful analysis, we establish at what level of complexity the problem belongs. Are we, for instance, dealing with a tool that must be redesigned, or are we dealing with a manufacturing method in which up to now this tool has been used, or should we rethink the product itself in relation to its ultimate purpose? Answers to questions like these do not yield to "seat-of-the-pants" examination.

A second area of investigation (unavoidably entwined with the previous one) is the historical perspective of the problem. All that we design is an extension of the human being (usually from generalization to specialization). While a high-fidelity system, for instance, may be loaded with associational values and carry a great deal of status, basically it is an extension of the human ear. As we have seen in our six-sided function complex (Chapter One), all design must fill a human need. The history of man's emphasizing or de-emphasizing particular needs and how they have been met is vital to the understanding and initiation of new products or systems serving these needs. Furthermore, such needs will be re-examined and regrouped with other needs or systems as the culture changes. Thus when the human, historical coordinates of an idea are found, certain principles can be

Experimental configuration of high-fidelity speakers, based on the dode-cahedron. "Ideal" sound cones happen to follow the continuation of planes extending the edges of a dodecahedron. This design uses twelve 93-cent speakers; two such speaker clusters give the stereo equivalent of a system costing ten times as much. Author's design.

applied to find out what particular phase of the idea we are dealing with.

Another consideration must be that of human factors. If we assume that all design is an extension of man (either good or bad), the relevance of humane values is obvious. Any design, on this level of consideration, is an organic substitution or implant (much like a transplanted heart, an artificial kidney, contact lenses, or a prosthetic hand). As such it must be recognizable and usable not only by the so-called "five senses," but also by the inner senses, both psychological and kinesthetic. Furthermore we must recognize the artificiality of this divorce between outer perceptions and the inner responses in man, for it gravely jeopardizes any unified human factors study.

Next, in integrated design we must attempt to place the problem in its social perspective. We have paid lip service to the concept that the entire factory system and automation (both are, as of this writing, supreme extensions of man) will result in making all that we feel we need available without effort, to all people, in all places, and at all times. But as our living patterns (and what we now think of as needs) change radically, the ultimate consumer values may no longer be "availability" and "effortlessness." Taking the long view, we can see that our attempts to remove all of our activities from the manual to the mechanical and then to the automatic indiscriminately may be quite wrong, as we have seen in examining the automobile vis-à-vis our "Triad of Limitations" in an earlier chapter. Chronically, we have failed to distinguish the means from the ends, and we have made mechanical what should have remained manual, and have made automatic that which might have been more rationally replaced with an entirely different system. A good example of such wasted energy is the automatic gear shift. The actual energy expended by the driver when shifting gears is incomparably smaller than the energy expended in manufacturing the automatic shift, not to mention the energy required to supply the factory and the automobile with the additional raw materials and man-hours required to make it. To quote Bob Malone (in an unpublished paper from 1957) on this:

Is the automatic gear shift then a true advance in humane design or not? Since it tends to *remove* man from
a basic and relatively simple use of his motor responses,
rather than to simplify and integrate the processes, we
can see that the validity of the automatic gear shift is
illusory. When a true need or desire is satisfied for a
passive human being without effort, the result is not
gratification, but rather a more complex level of dissatisfaction. The man caught helplessly in a natural catastro-
phe has good reasons to think about human dignity and
to wish that the material necessities of his life could
be met more simply.

Another level of social consideration in integrated design
must consider social groups, classes, and societies. Much
design must be re-examined to see how far it may perpetuate class systems and social status. As more and more
methods of social classification, stratification, and class
identity break down, there is a ready market for products
used to express social ambition and strivings for status.

Cassette-type tape recorders now come in nearly 40
models. All of these are battery-powered (with optional
power cords); they all use the same interchangeable tape
cassettes, are nearly identical in size and weight, possess
identical "guts" (often several different brand names are
manufactured in the same factory, like aspirin). . . . The
casing (or "skin") of these recorders is the identical black,
or sometimes gray, plastic; they have the same number of
control knobs; name plates are in identical positions, and
they all are sold in nearly identical black "leatherette" carrying cases. Nonetheless, they are priced from $22.95 to
$149.50 (with 15 price-breaks in between) and, what is
more surprising, are selling well in all price ranges. What is
the difference?

In the mid-range of these recorders (Wollensak, Philips-
Norelco, Concorde, DeJur-Amsco, etc., all selling for around
$70) the emphasis is on convenience, durability, ease of
maintenance, and repairability. The number of knobs is deliberately played down visually, to make the object look
"fool-proof." At the low range (Crown, Concorde, Sears Roebuck, etc., selling roughly in the $40 range) there is extra
chromium and aluminum, and the cases come in a range of

"decorator" colors. The emphasis here lies on "fun," bright colors, light weight, and its impermanent existence as a throw-away object. In the upper mid-range (V-M, Aiwa, Panasonic, etc., selling at about $110) some of the controls are of aluminum that has been anodized to look like copper or gold. Extra (non-working) control knobs exist on some of the models; various heraldic shields, emblems, and name plates have been added. The values stressed are those of social upward mobility, and there is cynically conscious catering to what the upper middle class considers "good taste." In the top range (Sony, Fisher, and others selling for around $130) there is great emphasis placed on "technical," computer-like appearance, a nearly complete absence of chrome parts and an uncluttered general appearance. The values stressed here are technical excellence, lack of ostentation, and simplicity.

This example should demonstrate how stylist-designers help to maintain present-day divisions between various income levels in a consumer society. By pooling present "state-of-the-art" knowledge regarding cassette-type tape recorders and manufacturing just *one* optimally useful type, it should be possible to reduce costs to a retail level of about $9, and

Trätofflor, still made in Ängelholm, Sweden, are a superb example of rational, vernacular design.

thus make possible a number of breakthroughs: magazines could be "published," letters "written," and education conducted—all on tapes that everybody could afford.

An opposite example is *trätofflor*, leather and wood slipper-shoes made in Ängelholm, Sweden. This footwear can be worn both at home and (with casual dress) on the street. They sell for about $4 a pair in Sweden. The upper part is made of cowhide; last and heel are shaped of wood. The soles are rubber. All three materials age well. These slipper-shoes are orthopedically beneficial to the foot as well as comfortable. They have a life expectancy of at least 4 years, can be worn in every kind of weather and, being nearly identical, cut completely across social and income classes, conveying no idea of status. (It is interesting to note in this connection that, of late, *trätofflor* are being made in a variety of textures, colors, and artificial materials. This makes them tend to wear out faster; repairs are more difficult and sometimes impossible.) They constitute, *in their original form,* a superb example of indigenous, non-manipulated design. Several brands of *trätofflor* have recently become popular in the United States where they became known as "Swedish clogs" and sold at higher prices.

Part of the philosophical and moral bankruptcy of universities and design schools lies in their ever-increasing trend to train students to become narrowly "vertical" specialists whereas the real need is for broad, "horizontal" generalists or synthesists. Nearly everything in today's university milieu militates against educating for general synthesis. Prerequisite courses, co-required courses, "required electives," and empire-building by deans and professors with their own vested interests at stake make education for a broader future nearly impossible. When we remember that the price which a species pays for specialization usually is extinction, this becomes even more criminal. George B. Leonard and I seem united in our feeling that most of what passes for "education" today is in reality a "crime against humanity" as defined by the Nuremberg Laws of 1945, and that most educators could be imprisoned for violating at least 6 of the 10 points of this same code for experimenting upon human beings—without their consent, without the subject's right to terminate experimentation, without the subject's right to change his conditions, for engaging in torture, etc., etc. It is to the credit of the young people of today that they have smelled all of this out and are trying to change it.

Ideally, of course, groups of concerned young people of all ages would meet together to engage in design. This would mean to learn, study, teach one another, experiment, engage in research and discussion, and interact with one another and with people from disciplines not generally subsumed under the heading of design. Such a group would be small (30 to 50 in number), and its members might stay together for weeks, months, or even years. Individual team members or small groups might, at will, detach themselves from the group, traveling or working directly with other groups or with manufacturing systems. Computer-assisted learning programs, as well as computer-assisted data acquisition, storage, and retrieval would of course be available to all members of the team.

But it is probably more meaningful to determine what can be done right now and in the immediate future.

In establishing a 5-year undergraduate curriculum for industrial and environmental design (at Purdue University), I took care that each student's program of study consisted of as free and broad a mix as was possible. We attempted to break down the false dividing lines between the various specialized fields of design such as visual design, interior design, industrial design, etc. Part of this was also training with twentieth-century tools of communication and expression such as computer sciences, photography, kinetics, cybernetics, electronics, and film-making. In addition to exploring verbal, visual, and technological methods of transmitting information, the students were encouraged to participate in other disciplines of concern to integrated comprehensive design. Thus sociology, anthropology, psychology (perception, human engineering factors, ergonomics), and, in fact, all the behavioral sciences were stressed. Because both individual human beings and social groups are biologically functional, the so-called life sciences must be a keystone in the study of systems, forms, structures, and processes. Hence, a study of chemistry, physics, statics, and dynamics was more than augmented by work in structural biology, ecology, and ethology. This led to courses in theoretical and applied bionics and biomechanics (cf.: Chapter Nine). Finally, nearly one third of all undergraduate time was left open for entirely free electives, which meant in practice that a student could assemble a "minor" in some area that was of concern to him such as anthropology or political science.

It is unfortunate that almost all schools or departments of design in the United States require an undergraduate degree in the same field as that in which the student hopes to do graduate work. We chose a different way, because of our passionate belief that the true design needs of the world must be carried out by cross-disciplinary teams. Hence, for graduate work we did not require 4 or 5 previous years of study in industrial design, architecture, or some other design area, but preferred taking our young people from the field of behavioral sciences. This added meaning to their work.

I make the (in today's educational circles, radical) assumption that my students are in class because they wish to study design. It is for this reason that we must reject the notion currently popular in America that the professor's role is authoritarian, disciplinary in nature, and that the teacher should be a part-time policeman. Therefore, students are free to come and go, and their class attendance is never noted down in an ominous little black book. Grades (a mechanical method for determining the relative proficiency of each student within the group) are never used in that manner. Instead, students whose ability is clearly superior at a given point in time are encouraged to "skip" entire semesters or even school years.

Those who seem to find it difficult to work meaningfully in design are counseled to go into other fields, or other schools. This relaxation of the professor's role as a disciplinarian removes one of the most harmful and destructive side effects of American education: competition and ag-

gression. It is plain to see why these twin drives are en
couraged in our educational processes, for without them the
profit system would fail utterly. Each student should be able
to demand the "highest standing in class" (a grade of A) for
a semester's work *before* the semester begins. This, and the
other grading procedures outlined above, fulfill a double
purpose. For one, they expose the bankruptcy of the com
petitive university system; secondly, they liberate the student
from the cares, worries, and pressures over his marks which
sometimes end in suicides. *Instead the student becomes a
participant in his own growth, is changed by his environment, and in turn, changes it.*

Today's student (child born in the era of television,
electronic information, and film) brings many different
skills to school before the first lesson has been taught. Unavoidably, he will be possessed of more recent, more accurate, or more relevant information in some fields than his
professors. Therefore, a class of 10 students and one professor is really a group of 11 teachers, 11 researchers in
search of knowledge, whose differing backgrounds comple-
ment each other. In schools where I have worked we encouraged the students to teach each other. If we were lucky
enough to have a student in class who worked in the electronics industry at some time or who drew exceedingly well,
he would be asked to take over the relevant teaching. For it
has become abundantly clear by now that the main task of
the school is to learn from and be changed by the students.

Students in advanced classes must have the right to vote
on who teaches. In my school they help us write our ever-
changing curriculum and frequently initiate entirely new
courses which they feel are needed. In order to experience
different working conditions, students not only work on individual projects but are also frequently given an opportunity to work in "buddy-teams" (two students). Often
larger teams are formed encompassing students and professors from different disciplines. The problems to be solved
may vary in time from simple 2-hour exercises to problems
lasting a month or two. In some cases, a larger team may
work on a more formidable problem for a full year. Since

each student, in order to learn the meaning of integrated comprehensive design, is encouraged to analyze each problem given thoroughly for social and human content, every student has the right to refuse work on any particular problem and to substitute a different problem of his own. Students also have the right to challenge whether a problem should be undertaken by the entire class. Such topics are settled in free and open discussion; from time to time problems are changed or substituted as a result of these discussions.

The old saying, "A teacher can learn from a student," happens to be literally true. It is for this reason that I encourage students to involve themselves in teaching in some manner. It is only through explaining the absurdities of our social system to others that we learn to see through them ourselves. And it is through teaching and working with groups usually not within the experience of the designer that we begin to understand the true needs of people or sometimes even the existence of certain groups. As Ho Ching-chi wrote in his introduction to his opera, *The White-Haired Girl:* "The people are our teachers, and it was they who taught us what work to do. They are our most reliable judges and authoritative critics. They are sometimes the creators of this art. . . ." (Peking, 1954).

For this reason, too, I urge students to travel widely and to work in many jobs—jobs not necessarily in design. They may work in offices, industry, or factories and farms. Such work forms a required part of their study during summer vacations; a full year of "internship" is helpful whenever practical.

A more ideal learning environment, possible today, is described in Chapter Twelve.

As mentioned above, the experience of working as part of multi-disciplinary teams is essential. This may be one of the hardest things of all to teach. Young designers have been sold and over-sold the concept of the lonely, struggling genius, the individual problem-solver. Reality does not bear this out. Most working designers today find themselves part of a team (like it or not). They may attempt desperately to hold on to the reassuring self-delusion that they are working alone, but, in fact, they are not. A typical marketing enterprise today will consist of upper management personnel, market and motivation research experts, advertising people, production engineers and, often as not, consumer psychologists. Some of these people wield decision-making powers; others may work as consultants; still others may have broad advisory powers. In many cases, in fact, the designer finds himself just a sort of vermiform appendix to the marketing-cum-advertising brigade.

Integrated design needs teams of specialists too—specialists from areas whose orientation is not private profit-making, but rather a human and humane concern for man and his environment. Such a team might consist of a designer, an anthropologist, a sociologist, and people in specialized areas of engineering. A biologist (or at least someone versed in bionics and biomechanics) and medical and psychological experts would complete the team. Last, but very importantly, the people for whom the design team works must have representation on the team itself. Without the cooperation of the eventual "clients," no socially meaningful design can be done. When students are first faced with this concept, they try to escape a confrontation with some client groups by assuming that there will be communication difficulties, or that the members of the group might be too ignorant to realize their own needs fully. Such lack of faith in the people can never be really justified.

I have operated on design teams which included uneducated rural poor people, small children, or mentally disturbed patients. While a communications link was slow and difficult in coming about, we finally succeeded in every case, and as a result were made directly aware of needs which the professional opinion-takers didn't realize or considered unimportant.

The above is by no means a "blueprint" for an ideal team, as many other disciplines may need to be represented in specific cases. Aside from the socially progressive orientation of such a team (compared with the marketing team mentioned earlier), our new team no longer consists of

LEFT: Chair desiged for an international competition. Weight is relieved from the spinal column and distributed over the fatty tissues of the back. Designed by author, while still a student. The chair sold successfully, but was eventually withdrawn from the market by the designer on the grounds that it was ugly and expensive.

BELOW: The chair redesigned for easy manufacture through cottage industries in the Southern Appalachians. It is simpler and less expensive, and money goes directly to the people who carve it. Author's design.

Candlesticks specifically designed to be made through home-cottage industries in Southern Appalachia where these candles are also made. Author's design.

managerial decision-makers and consultants, all of whom think of the designer as a sort of glorified "errand boy." Rather, it is a free and equal team of professional experts whose only aim lies in designing-planning. It is the task of the design team not only to *solve* problems, but also to search for, isolate, and identify problems that need solutions.

It is in this last area: locating, isolating, and identifying problems, that the schools fall lamentably short and in fact often provide no practice for the student at all. Students in most learning situations are asked to solve projects. This means that a "special-case" situation is presented to the student and, after a certain amount of time, the student is expected to regurgitate a "special-case" answer to the teacher. Thus, he may be asked to make a ceramic teapot for 6 cups of tea, and this (embellished in his own particular way) is precisely what he will return to the teacher. Instead of the concept of a ceramic teapot, we could as easily have substituted the design of a better chair, a city plan for the Chicago ghetto, or a magazine cover. It really does not matter what specific design problem gets plugged in, for in each instance it is a "special-case" situation, and that is *not* the way things work. Even if *all* the problems assigned were to be socially relevant, the "general-case" learning experience of the student would still be nil. The human mind, as well

as human problems, continuously moves from generalizations to particulars and then broadens out to generalizations
again. It is a never-ending pendulum swing between "special
case" and "general case."

A problem can be assigned as either a special or a general
case. What is important is the functional processing of the
idea by the student, the designer, the team, or the class, as
well as their understanding of this process and its links to
other similar processes. A few examples will suffice: a problem may be assigned as "special case"; for instance: "Design
a chair!" The student will then move from this special case
out towards the generalization "chair." He will review alternative design strategies and, from these, develop a number
of so-called "sets." These "sets" are various directions, general and often mutually exclusive, in which the problem can
be solved. Just a few of these "sets," which the student may
discover in a general case, could include: a disposable chair,
a chair for people with injured backs, a chair for children in
primary schools, a method of sitting in a boat, a chair for
performing some specific technical task such as playing in
a string quartet, a "fun" chair that will appeal to a particular sub-group, etc., etc., etc. The student now selects his
particular "set" out of the general case and proceeds to work
towards his own special-case solution. This is shown sche-
matically in Diagram A.

Diagram A:
One design "Event"
Special case to general case to
special case.

A "general-case" problem statement might be: "Design
something to help underdeveloped countries!" The student
now has to engage in a great deal of research from various
sources and disciplines. From these he may eventually narrow down to the special-case concept "bicycle-like power
source." But in developing this design concept he will unavoidably find many spin-offs and spill-overs and thus again

arrive at many general-case solutions and applications. (It
is specifically this type of problem that is almost never set
in school.) This process (looking somewhat like a butterfly
or a bow tie) is illustrated in Diagram B.

Diagram B:
One design "Event"
General case to special case to
general case.
(Or Team problem)

It should be obvious that in any team design problem the
flow diagram will be as in Diagram B. Many different students are assembling general-case information through research and bringing this together as an information package
to be commonly shared under "special case." From here they
will again fan out to many general-case solutions.

It will be useful to remember that both Diagram A and
Diagram B can be thought of as single links in continuous,

Diagram C:
A series of design
"Events", cyclic
in nature.

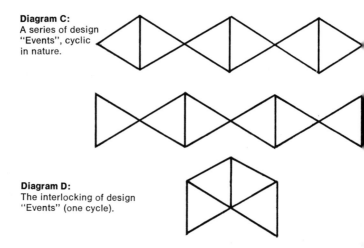

Diagram D:
The interlocking of design
"Events" (one cycle).

cyclic chains, as in Diagram C. As shown in Diagram D,
both "events" (Diagrams A or B) interlock.

A series of possible design "events" (Diagrams A *and* B)
will yield an omni-directional, 2-dimensional net of equilateral triangles organized into close-packed hexagons, with
no wasted space. This is shown in Diagram E.

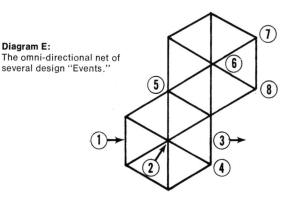

Diagram E:
The omni-directional net of
several design "Events."

By studying the schematic function of Diagram E, its use
can be seen easily. The designer or student may start with a
general-case idea input at No. 1, confidently expect to reach
special case at No. 2, and hopefully plan to derive an answer
at No. 3. However, No. 2 is a locus for at least 6 different
disciplines, and he may in fact eventually emerge at either
general- or special-case points No. 4, 5, 6, 7, 8, . . . or "n."
Diagram E then becomes a schematic representation of a
series of interlinking "events," each one of which can be
represented by a flow chart, each flow chart carrying with
itself the bias or "set" of its own particular discipline.

(Note: We must remember that our schematic representation of the design process of multi-disciplinary teams as
shown in Diagram E has been reduced to a 2-dimensional
sketch. A truer representation of this flow of information
and processes would be a 3-dimensional model consisting
of a number of tetrakaidecahedra, close-packed in space.
Loci or information-exchange areas would then be represented by the hexagonal faces; directions of design processing by the axes of the square faces.)

Now let us examine the flow of a real design problem
through our schematic. *An illustration of this will be found
as Diagram F:* At No. 1 (triangle "a") the designer enters
the picture with a special case problem: "Design a chair."
Triangle "a" represents his normal data-gathering phase,

Diagram F:
Schematic representation of the behavior of a multi-disciplinary team. Only a small section of the hexagonal net is shown.

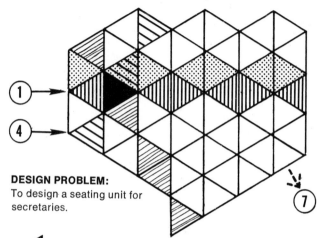

DESIGN PROBLEM:
To design a seating unit for secretaries.

 Designer's cyclic path (triangles a, b, c, d, e, etc.) if uninterrupted by other disciplines.

 Doctor's cyclic path (triangles u, v, w, x, y, z): the treatment of work-induced diseases.

 Sociologist's cyclic path (triangles p, q, r, s): work habits and attitudes of secretaries in offices.

 "Client" group's cyclic path (triangles g, h, i, j, k): in *this* case, office secretaries, doing their work.

 Intersection of various disciplines' paths.

 Cyclic paths of other groups not concerned in this particular problem.

1: Entry point of designer.

4: Interface between sociologist's path (near point of his emergence) and consultation with some other discipline, say, engineering.

7: One of many possible, unpredictable emergence points by the team.

bringing him to point No. 2, the general-case collection of his ideas. At this point he is still independently acting as a designer; if left to his own resources he would eventually emerge at No. 3 (triangle "b") with, say, a low-cost desk chair for secretaries. Still left to his own resources, he might now, still at No. 3, begin his next design job (another chair, a tool, or whatever). This would carry him through triangles "c" and "d." (In fact the undisturbed activity of a typical designer-specialist of today can be read as the cyclic axis: a, b, c, d, e, etc.) However, our designer friend is not a specialist, but rather a member of a multi-disciplinary team. When he reaches No. 2, he has not only reached general case data, but also the intersection of several other lines of thought. For here the medical doctor, for instance, will bring forward information regarding sitting postures (normally the doctor's own cyclic axis would continue toward triangle "w" as well as x, y, z—the treatment of work-induced diseases). Here at No. 2 the sociologist (axis: p, q, r, s,) and some secretaries as representatives of the client group (axis: g, h, i, j) also intersect. Our designer, through meeting and working with many other team members, may finally emerge at, say, No. 7 (triangle "m"), which might be a systems design for a communications device that permits secretaries to work in their homes.

As has been explained earlier, in order to understand all the ramifications of integrated comprehensive design fully, it is necessary to try to become aware of all the parameters that have bearing on the design process. Since there are so many factors and variables involved (more than can possibly be kept in mind), I find that the simplest solution is to *externalize* it by constructing a flow chart. A flow chart (as my students and I use it) is generally a large roll of brown wrapping paper pinned across an entire wall. Written down on it are all the various aspects that have bearing on analyzing the design.

Recently, during the primary design stages of a playground for a slum area, such a chart was constructed. Some of the factors that appeared on the chart were: Psychological and physiological needs for participation, exercise, and group-needs of children at various age levels. What kind of supervisory personnel would be needed, and how available they were in the area. What kinds of playground equipment could be designed and built and with what resources, what tools and processes. How money could be raised for this.

What materials could be used for constructing equipment and toys, and what were the characteristics of these materials under: (a) extremely hard wear and use; (b) frost, ice, snow, storms, and hard rain; (c) prolonged use over a period of 5–15 years; (d) dangers of shearing, splintering, torque, or fracture while being used by a child; (e) toxic characteristics of the various materials and coloring agents; (f) perceptual and psychological responses of children (at various age levels) to the colors used; (g) relative ease of care, maintenance, repair, and replacement of equipment, etc. We also included questions regarding the setting of the playground within the neighborhood area with such determinants as: (a) location of playground entrances in relation to main traffic arteries; (b) number of streets to be crossed by children hoping to use the playground; (c) illumination of the playground at night; (d) accessibility to homes and other neighborhood centers such as nursery schools, kindergartens, day-care centers, etc.

We also listed possible ancillary services such as: toilets, drinking fountains, a swimming pool, a wading pool for small children, telephone facilities, first aid equipment, a rain shelter, benches for older people, landscaping (grass-planting, bushes, trees, and flowers), etc. We also listed activities other than play which might take place within this area such as outdoor concerts, motion picture showings, or street theater for older people; "story-time" and "sing-alongs" for smaller children; dances and athletics for teen-agers, etc. Climatic considerations also had to be applied: could parts of the playground be flooded for ice skating during the winter? Could some of the hills (which we were to create with bulldozers) be used for bobsleds, sleds, and skis? What about drainage problems during rainstorms and after the melting of the permafrost in the spring? These are just a few of the areas which we considered on our flow chart.

A flow chart works in a quite simple way: We listed all

he parameters we could think of (some of which are mentioned above), putting each under whatever classification eemed to make most sense. Under activities, for instance, we might list: climbing, jumping, running, sliding, singing, alking, and many, many more. After everything had been isted, we then began to establish *relationships* where none eemed to exist before. For example: under "materials" we isted sail cloth or heavy canvas. Its characteristics are (when stretched and supported like a membrane) buoyancy and comparatively resilient softness. This could now be brought into a direct relationship with "jumping" and suggest a trampoline-like structure to us. One of the most important functions of a flow chart is that new relationships or inter-linkages can be read directly off the wall and that solutions, or at least directions for solutions, emerge without their ever having been consciously listed. Another point about a flow chart, of course, is the fact that it can, by definition, *never be complete*. That is, new concepts and entire new categories can be added almost indefinitely, and hence new relationships and inter-linkages will constantly emerge.

At this point, half of the flow chart (or triangle "a" in Diagram A, above) has been completed. The second half of the flow chart (triangle "b") will consist of implementation. That is, who does what, when, how, and by what date. Here again, alterations and additions can be continuously performed. The entire design team keeps the flow chart on the wall until *after* the design job is completed.

We can now establish the work flow of any design job:

(1) Assembling a design team representing all relevant disciplines, as well as members of the "client group."

(2) Establishment of a primary flow chart (triangle "a" part only).

(3) Research and fact-finding phase.

(4) Completion of the first half of the flow chart (triangle "a").

(5) Establishment of the second half of the flow chart (triangle "b"): "what to do."

(6) Individual or "buddy-team" or team design and development of ideas.

(7) Checking of these designs against the goals estab-

lished in the flow chart, and correcting both th
designs and the flow chart in the light of these de
sign experiences.

(8) Building of models, prototypes, test models, an
working models.

(9) Testing of these by the relevant user-group.

(10) Results of these tests are now fed back into the flow
chart.

(11) Redesign, retesting, and completion of the desigr
job, together with whatever written reports, graphi
communication, statistical support data, or workin;
drawings are necessary.

(12) The flow chart is then preserved, to be used as
follow-up guide in checking actual in-use perform
ance characteristics of the designed objects. Afte
this the flow chart is filed, to be used as a guide fo
future design jobs that are similar in nature.

It should be obvious that in reality the design process car
never follow a path quite as linear and sequential as sug

Right-handed half of the flow-chart (part "b"). This is a stud
for using old railroad cars to bring health-care services to poo
rural areas. Graduate student-team designed by Jules Belanger
Pierre Bossé, David Koropkin, and Louis Noriega, Californi
Institute of the Arts.

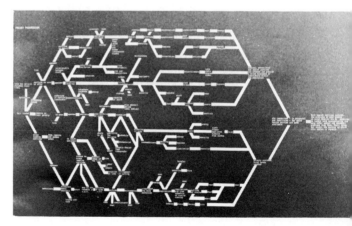

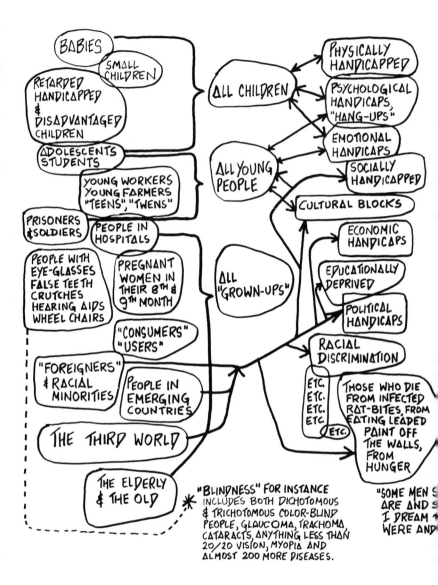

BABIES

SMALL CHILDREN

RETARDED HANDICAPPED & DISADVANTAGED CHILDREN

ADOLESCENTS STUDENTS

YOUNG WORKERS YOUNG FARMERS "TEENS", "TWENS"

PRISONERS & SOLDIERS

PEOPLE IN HOSPITALS

PEOPLE WITH EYE-GLASSES FALSE TEETH CRUTCHES HEARING AIDS WHEEL CHAIRS

PREGNANT WOMEN IN THEIR 8TH & 9TH MONTH

"CONSUMERS" "USERS"

"FOREIGNERS" & RACIAL MINORITIES

PEOPLE IN EMERGING COUNTRIES

THE THIRD WORLD

THE ELDERLY & THE OLD

ALL CHILDREN

ALL YOUNG PEOPLE

ALL "GROWN-UPS"

PHYSICALLY HANDICAPPED

PSYCHOLOGICAL HANDICAPS, "HANG-UPS"

EMOTIONAL HANDICAPS

SOCIALLY HANDICAPPED

CULTURAL BLOCKS

ECONOMIC HANDICAPS

EDUCATIONALLY DEPRIVED

POLITICAL HANDICAPS

RACIAL DISCRIMINATION

ETC. ETC. ETC. ETC.

ETC.

THOSE WHO DIE FROM INFECTED RAT-BITES, FROM EATING LEADED PAINT OFF THE WALLS, FROM HUNGER

* "BLINDNESS" FOR INSTANCE INCLUDES BOTH DICHOTOMOUS & TRICHOTOMOUS COLOR-BLIND PEOPLE, GLAUCOMA, TRACHOMA, CATARACTS, ANYTHING LESS THAN 20/20 VISION, MYOPIA AND ALMOST 200 MORE DISEASES.

"SOME MEN S ARE AND S I DREAM WERE AND

OM THOUGHTS:

LL PEOPLE ARE HANDICAPPED
OR AT LEAST PART OF THEIR
VES, IN SOME WAY. AS DESIGNERS
E MUST FIND THEIR <u>REAL</u> NEEDS.

OTHER SOCIETIES
HE PAST? HOW ARE
IETIES LIVING IN
RTS OF THE WORLD?
PROBLEM ALREADY
ED, SOMEWHERE OR
N ELSE? "GOOD TASTE" IS AN
INVENTION OF MERCHANTS,
DESIGN & FOLK-ART DON'T
HAVE IT. "GOOD TASTE"IS
TO CREATE FASHION FOR
tees & PROFIT.
, creative
loving is
t than
n construct
models of
ent environ-
them out.

"YOU HAVE TO MAKE UP YOUR
MIND EITHER TO MAKE SENSE
OR TO MAKE MONEY, IF YOU
WANT TO BE A DESIGNER."
BUCKY FULLER

JUST DO THINGS AS WELL AS WE CAN"
BALINESE PROVERB

BUILD LIFE FROM REFRIGERATORS,
EDIT STATEMENTS AND CROSSWORD
T IS IMPOSSIBLE. NOR CAN ONE EXIST
GTH OF TIME WITHOUT POETRY, WITHOUT
UT LOVE."
ANTOINE DE ST. EXUPERY

Y
WHITE,
VE
Y SINGLE
KIND HAS
LIPSTICK FOR
BUT TO CREATE DEODORANT FOR HER PIMP IS ANOTHER.

"TELLING LIES TO THE YOUNG IS WRONG
PROVING TO THEM THAT LIES ARE TRUE
IS WRONG
...THE YOUNG KNOW WHAT YOU MEAN.
THE YOUNG ARE PEOPLE.
TELL THEM THE DIFFICULTIES CAN'T BE
COUNTED
AND LET THEM SEE NOT ONLY WHAT WILL BE
BUT SEE WITH CLARITY THESE PRESENT
TIMES."
YEVGENY YEVTUSHENKO

"If a free society cannot help
the many who are poor, it
cannot save the few who are
rich." JFK

"WE ARE NOT REHEARSING FINAL GESTURES, WE WANT
LIFE AND WE SHALL DEFEND IT." CHE

"WHEN YOU MAKE A THING, A THING THAT IS NEW,
IT IS SO COMPLICATED TO MAKE IT
THAT IT IS BOUND TO BE UGLY.
BUT THOSE THAT DO IT AFTER YOU
THEY DON'T HAVE TO WORRY ABOUT MAKING IT.
AND THEY CAN MAKE IT PRETTY, AND SO EVERYONE
CAN LIKE IT,
WHEN THE OTHERS
MAKE IT AFTER YOU." PICASSO

"IT IS NOT ENOUGH TO ALLOW DISSENT.
WE MUST DEMAND IT. FOR THERE IS MUCH
TO DISSENT FROM..." RFK

THE ONLY THING THAT DOES NOT CHANGE
IS CHANGE. EDUCATION MUST BE ECSTATIC
AND FOR TOMORROW...

gested by this example. (For one thing, new research data emerges continuously.) Nonetheless, this should give some idea of a flow chart and its use in design.

While participating in a design conference held by the Scandinavian Student Design Organization (sdo) at Copenhagen in the summer of 1969, it was my job to construct a "general case" portion of a flow chart, concerned with the social and moral responsibility of the designer and his position in a profit-oriented society. This is a large job indeed. In fact, this entire book attempts to address itself to precisely that question. Nonetheless, the flow chart is reproduced in this book, and a few explanations are in order. As the meeting dealt with the question of handicapped people, columns A and B attempt to point out that *all* people are handicapped in some way or another for at least part of their lives. One small entry in column A has been isolated: the concept "blindness." Here an attempt has been made to show that even people wearing only mildly corrective glasses are somewhat handicapped in the area of seeing. Seven out of more than 200 optical disturbances have been listed.

Column C lists the *real* needs of the people (partially and imperfectly at best). Column D, "What the people want," is empty for reasons that will appear later. Column E lists what the people are *told* they need and want—in other words, the substitutions made by our society for real needs. Column F shows the processes used in achieving these false goals, and column G attempts to show just a few of the repressive measures used by our society to prevent the attainment of *real* goals. Column H, labeled "How to change this," lists the revolutionary and evolutionary activities of education, creativity, social planning, and research. Through links it attempts to show that it is these processes that are "design." Under column I a number of thoughts (even catch-phrases and slogans) have been put down, in the hope that more will be added. Interestingly enough, it is the connection between social planning, revolution, creativity, education, evolution, and research under column H and these random thoughts under column I that provide us with our first major insight into column D ("What the people want").

Finally, column K attempts to show the make-up of a design team, together with some of the specific disciplines that must be fed into it. At the extreme right of the chart, a large arrow points to 6 activities. It is these 6 activities that will eventually make up the right half (triangle "b") of the flow chart, things to do, the "operative" phase. *All socially relevant, integrated comprehensive design must be operative—that is, related to the true needs of the people.*

The accompanying flow chart is far from complete, both in the number of entries or even in the number of relationships or linkages established between what entries have been provided. The reader is encouraged to play with the flow chart, to add to it and discover his own relationships. The right half has been purposely left open: completed, it will form a social and political blueprint for tomorrow—for society as well as for design—far beyond the scope of this book.

It can be argued that the subject of this particular flow chart is too broad. But flow charts are, by their very nature, "general case" statements. A very narrow subject would have become too highly technical for general understanding.

One of the most dangerous aspects of "higher education" in design is its constant reiteration that the function of the school is to make it possible for the student to get a job and, once he or she gets it, to hold on to it for dear life. This has turned design education into the narrow field of vocational training. Most young people today have seen through this cynical method of perpetuating a bankrupt system. Most of them no longer worry about coming to school to learn a skill to make a lot of money. They *know* that they will make money, and they are concerned only with *how* they will make that money. In other words, they are less concerned with their own "standard of living" than with everybody's quality of life. When some of them show concern over money-making, they merely reflect the projected fears and insecurities of their parents, their teachers, and society at large. It is wrong to tell young people that the most important thing is to find a job. It matters what *kind* of job it is. And it is far better for a young designer to design a few socially valid things in the evening and wash dishes for a few months during the day than to put his skills and talent at the disposal of a wrongly directed system destructive to human dignity.

Some of my own students have had difficulties in getting a first job that was not a compromise of their convictions. Some of them have consciously turned down well-paying, status-conferring employment opportunities in industry in order to seek and find work in clinics, hospitals, or under-developed regions for less pay. This has been a little hard on them, but it is getting easier each year. The students starting to study design today will have little or no diffi-culty doing worthwhile work when they graduate in the mid-seventies. For in a world choked with glittering gadgets which at the same time does nothing for peoples' true needs, different tasks will become most important. And, at any rate, design is never a job; it is a way of looking at the world and changing it. The second dangerous premise which most schools and industry tell to the young is that to be a de-signer confers "instant status"; designing is a fashionable profession. This lie must also be exploded.

All this, however, is merely a broad philosophical back-ground. What about specifics? What sort of things can be used in training students to design?

During the summer of 1968 a multi-disciplinary team of design students (under the guidance of Yrjö Sotamaa, Zol-tan Popovic, Barbro Kulvik-Siltavuori, and Jorma Vennola) worked with me on a small island in Finland and invented, designed, and built a foldable, moveable environment for children with cerebral palsy. This environment included toys, exercising devices, and many other pieces of equip-ment. We met in Helsinki, after the 8 student members of the team had already played with and interviewed the chil-dren. They had also spoken to parents, visited clinics, play-grounds, and homes. They had found that little or no equip-ment had been specifically invented or provided for children with cerebral palsy and that some of the toys now in use to train such children in specific motor skills were inhumane

An exercising and play environment built on Suomenlinna in Finland. Designed and built by a cross-disciplinary team of students under the direction of Zoltan Popovic, Yrjö Sotamaa, and Victor Papanek.

and barbaric. (CP children must be trained to use their thumb and index finger in grasping. It is their natural tendency to use the other three fingers instead. Until now they have been trained by strapping or tying these three fingers together, thus being forced to use thumb and index finger alone. Several toys were designed and made that provided reward-sequences and enjoyment to the child only when he used the first finger and thumb. In this way the medieval practice of forced restraints could be abandoned.) The students also found that clinics and hospitals were drab and unexciting.

We made a flow chart and met as a team, together with two experts in child psychology and neurophysiology from Sweden. As a team we spent a total of 12 hours developing a 2-meter cube which knocked down into two sections each 2 × 2 × 1 meters in size. This module permits the 2 parts of the cube to be moved easily from clinic to clinic, to be carried through doors and transported on small trucks. Once erected at the clinic (indoors or outdoors), the cube unfolds into a play environment that is 2 meters high, with equipment covering an area of 16 square meters. It is bright and colorful and includes slides, climbing, a crawling surface, and many individual toys. It is also easy to build and low in cost. Our first prototype cube was built and completed (including toys) in 30 hours of teamwork and then tested with children. We called it "CP–1" to suggest that it was merely the first of a generation of similar cubes, each one of which would be modified by testing and experiences with children. We also assumed that other cubes (for instance, for hydrotherapy, autistic and retarded children, etc.) will eventually be built. A fuller discussion of this, together with photographs, will be found in *Industrial Design*, November, 1968.

During January, 1969, students at the State School of Design at Oslo, Norway, worked with me for 2 weeks to develop a playground-environment for the backyard of a group of old, inner-city apartment houses. The 6 buildings comprised in the area housed nearly 70 children who could play only in the dangerous streets as the 3 backyards were given over to garbage pails, high metal fences, and laundry

ABOVE: An exercising vehicle for disabled or retarded children. Designed by Robert Worrell, as a student at Purdue University.

BELOW: Vehicle for exercising children with weak arms and shoulders. Designed by Charles Schreiner as a student at Purdue University.

A least-effort vehicle designed for children with cerebral palsy. Both the pedals and the arm move the vehicle. Whatever limb the child can use propels him, while the other limbs are exercised. Designed by Charles Lanius, as a student at Purdue University.

lines. The students began interviewing the residents of the various buildings.

Through their interviews many new facts emerged for our flow chart: Elderly people declared themselves uninterested in meeting new people or talking with them, but took great pains to invite our students into their homes for periods up to 2 hours and served them tea and cookies while telling them so! With some justice, we felt that some of these older people were unaware of their own motivations and needs, and could be drawn into the social group. Some might even be willing to help supervise play. Younger people were strongly in favor of constructing the playground, and many offered to help with money; a few ashamedly confessed that they were too poor to help financially. We felt that it was precisely the poorer people who, by being taught to help through their work instead of monetary contributions, could be drawn more strongly into a social group engagement.

The parents promised to support the work, as the nearest park or playground was several miles away. The youngest person interviewed was a four-year-old boy who immediately began peering out the window to see if the playground was there yet. We can imagine that he kept pestering his mother and the neighbors day after day, relentlessly, as small children do, and that, in fact, he was probably our strongest propaganda weapon. After our flow chart had reflected such diverse data as interview results, mean annual sunshine distribution, newer methods of storing the garbage, etc., we were ready to go to work.

The students were appalled to find that the backyard was infested by rats and that the children played with the rats and thought of them as pet animals, something along the order of small dogs. We saw that design would have to go beyond a playground to include factors of public health and hygiene. Because of the social relevance of this project, other students from the Architectural School, the School of Landscape, and Oslo University became interested and volunteered their help, even though students from these schools normally have little or no contact with the State School of Design.

I must admit that at first many of the students became interested because of the novelty of the problem. Later they found that being involved in this type of social design is much more difficult than creating still another teapot or a perfect salt cellar. Many were discouraged, and some dropped out. Yet the landscaping, the design and development of equipment, drawings, and a 3-dimensional model of the eventual playground were completed in time.

The next step would be to invite the people now living in these tenements (including the childless and the elderly) to see the work that the students had done. It would then become possible to mobilize the people living there to work with the students in building the play environment. Benches and "quiet corners" had been provided for older people; an inner courtyard apartment was to be turned into a laundry area with one or two washing machines so that the mothers could wash their clothes, drink coffee, gossip, and watch the children. But even if this playground were to be completed, this would not end the work.

By accepting the responsibility of involving themselves with the people living there, the students had also accepted a more permanent responsibility towards these same people. It was up to the students that outdoor movie showings, guerilla theater, poetry readings, and "sing-ins" be brought to the backyard on long summer evenings. Through engaging in these activities, the students came to a closer and "operative" understanding of the people's problems; the people, in turn, assumed a more active role in shaping their own future and gained pride and identity. A secondary result should spring from the playground: its evident success should stir other communities and tenement blocks into similar actions.

At Purdue University we concerned ourselves with the problems of paraplegic, quadriplegic, spastic, and palsied children. We designed and built a series of vehicles with built-in motivational factors that will provide healthy exercise and training for these children. A study showed that the nature and extent of their handicaps and abilities varies greatly. Some can use only their arms, some only their legs; with others the entire right or left side is useless; a few have the use of only one limb. However, in many of these cases it is healthy to exercise the useless limbs. One thing which all these children have in common is a great enjoyment of speed. The vehicles illustrated were designed so that they can be operated with one or more limbs; the others are exercised in the process. The harder the child exercises, the faster he goes. Hence, enjoyment and exercise go hand-in-hand. The vehicles were tested with handicapped children and were turned over to local clinics.

Let me give another example of design used effectively for disadvantaged people.

The American South and the Midwest are criss-crossed by railroad tracks, with the depot usually in the downtown area of the town or village. So as one of our projects my students and I designed 3 trains, each to serve a specific function, to be staffed by trained personnel, and to be parked in a small town for 3 or 4 months at a time. One of the 3-car trains will operate as a vocational re-education center; the others will provide birth control information, ophthalmology clinics to issue eyeglasses, clinics to perform dental services, issue corrective devices for birth defects, and operate in disaster areas and as epidemic control centers.

One more project: One of our graduate students designed a series of pill containers that are child-proof and may save the lives of the many children who die each year in the United States alone from eating aspirin, tranquilizers, or sleeping pills, thinking that they are candy.

Our work for Mexican migrant farm laborers and for the white, alienated rural poor of eastern Tennessee and western North Carolina was desperately needed by them.

Because Navaho, Hopi, Zuñi, Mescalero Apaches, and other Indians feel that, for moral and religious reasons, they must have a large share in the construction of their homes and because, furthermore, they move twice a year, from their winter camp to their summer grazing land and back, and because they feel it is wrong to kill vermin, we worked on a shelter design for their very special needs. One of our graduate students developed a minimal shelter (related more closely to the Navaho's concept of space than to an "adapted" white man's house) that keys in exactly with the Indian style of life. The shelter is woven in parts (weaving is a major Navaho skill), and the wool is vermin-proofed during the weaving process. These woven parts are stuffed with dried native grasses that insulate against extremes of heat and cold. A number of these parts can be zipped together (thus the shelter is self-erecting) to form an enclosure of nearly any desired size. Zipped apart, it can be easily moved on horseback during the two annual treks. All in all, it is far better suited to the territorial, cultural, and religious aspirations of the Indians than the confections doodled up by the Department of Health, Education, and Welfare.

In ending this brief list of a few gadgets and designs, we should also concern ourselves with what the student has gained. Obviously he has engaged in research, worked with a team, met the needs of people, operated with a flow chart, and gained new skills and new insights. But the actual learning content of these problems is far greater than that, going from the immediate to the more permanent. A series of educational steps and learning experiences have taken place, all of them on an interactive level. It might be best to list them now:

(1) The student has located, identified, and isolated a problem. In so doing he has interacted with other members of a multi-disciplinary team and engaged in a meaningful work experience with a group of clients whose existence and needs were previously unknown to him.

(2) Through his work he has made the client people aware of the promise that design (applied intelligently) can hold out for them. He has satisfied their needs at least partially.

(3) By working with and helping the group, he has exposed

a) the needs of the group to society.

b) the lack of knowledge on the part of society regarding the needs of the group, or the very existence of the group.

c) the cynical indifference of the governmental power structure and industry to most of the genuine needs of people.

d) the inability of traditional design-as-it-is-taught to cope with genuine social problems.

e) the existence of methods and disciplines to work intelligently for these needs.

f) the obvious lack of schooling and training in this, the most important area of design.

(4) He has engaged in satisfying work; never again will it be possible for him to engage in the kind of design directed towards "good taste." Having experienced this kind of work, he will forever after feel a little ashamed when he designs a pretty, sexy toaster.

He will forever after feel a little ashamed when he designs a pretty, sexy toaster. . . .

12

DESIGN FOR SURVIVAL

AND SURVIVAL

THROUGH DESIGN:

What Can We Do?

Some men see things as they are and say, why?
I dream things that never were and say, why not?
ROBERT F. KENNEDY

AGAIN: Design is basic to all human activities. The planning and patterning of any act towards a desired, foreseeable end constitutes a design process. Any attempt to separate design, to make it a thing-by-itself, works counter to the inherent value of design as the primary, underlying matrix of life.

Integrated design is comprehensive: it attempts to take into consideration all the factors and modulations necessary to a decision-making process. Integrated, comprehensive design is anticipatory. It attempts to see trends-as-a-whole and continuously to extrapolate from established data and intrapolate from the scenarios of the future which it constructs. Integrated, comprehensive, anticipatory design is the act of planning and shaping carried on across all the various disciplines, an act continuously carried on at interfaces. In metallurgy it is at the boundary layers (the interfaces between crystals in metals) that action takes place under force. These very imperfections make it possible for us to shape and deform metals mechanically. Geologists tell us that the great changes on earth take place where forces meet along boundary lines. Here surf meets shore, fault blocks move in different directions. Diamond cutters cut along flaw lines, the sculptor's chisel follows the grain, and naturalists study the edge of the forest meeting the meadow. The architect's main concern is with the juncture between building and ground; the industrial designer is concerned with the smooth translation between working edge and tool handle, as well as with the second interface, the "fit" of tool and hand. Passengers relax visibly after that split second when the airplane finally leaves the ground, and for every navigational map of the ocean, there must be a thousand showing reefs and shorelines. We fight our wars over symbolic boundaries which we draw across our maps, and find life's most shatteringly poignant experiences are crossing the boundary lines of birth and death; our apotheosis is the sex act: ultimate encounter between interfaces.

It is at the border between different techniques or disciplines that most new discoveries are made, most action is inaugurated. It is when two differing areas of knowledge are forcefully brought in contact with one another that, as we have seen in a previous chapter on bionics, a new science may come into being. Frederick J. Teggart, the historian, says that "the great advances of mankind have been due, not to the mere aggregation, assemblage or acquisition of disparate ideas, but to the emergence of a certain type of mental activity which is set up by the opposition of different idea systems."

Acceleration, change, and the acceleration of change itself arise from the meeting of structures or systems along their edges. Intuitively, young people today have sensed this; their repeated use of "confrontations" is a symbolic, externalized illustration of this fact.

By its very nature the design team thrives on such confrontations, being itself born of interfaces. The design team is structured to bring many different disciplines to bear upon the problems that need solving, as well as to search for problems that need to be rethought. Its task is to do research to find our true needs and to reshape environments, tools, and the way in which we think about them.

Currently, it is fashionable to be concerned over the advent of the computer age. And although the foreseeably increased use of computers divides people into two sharply opposed camps, it is often conveniently overlooked that the viewpoints of both factions are essentially negative. The first group sees computers as a threat to organized labor, to the standard 40-hour work week, and to the Puritan work ethic, and finds much to ponder and fear in that. The other group, while realizing that the computer may finally help to phase out drudgery and back-breaking work, as well as that work's equivalence in monotonous, routine intellectual labor, also takes a negative view of the future. Here the threat seems to be mass leisure. The anecdote in Chapter Three about Piet Mondrian, painting as though he were a computer, illustrates the particular fear that artists evince when faced by data-processing machines.

But, as the old cliché has it, "Nature abhors a vacuum." As computers begin to take over (or as we relinquish to them) a greater share of those activities that we have heretofore thought of as exclusively intellectual—but which in fact are sheer monotony—new areas of engagement cannot fail to emerge. It is precisely here, at the juncture between computerized "work" and human "leisure," that the design team is located.

In a world in which agricultural and industrial work increasingly will be done through automated factories and in which most routine supervision, control, and computation is performed by computers, the work of the design team (research, social planning, creative innovation) *is the only meaningful and at the same time crucial activity left to man.* Inescapably, it will become the job of designers to help set goals for all of society.

Social historians tell us that the predicament of twentieth-century man can be traced unerringly to the discoveries of 5 men: Copernicus, Malthus, Darwin, Marx, and Freud. But during just the last 5 to 10 years the interfaces between sociology and biology, between psychology and anthropology, between archaeology and medicine have generated wide new insights into the human condition. Ten new books

—Robert Ardrey's *The Territorial Imperative,* Nigel Calder's *The Environment Game,* Edward T. Hall's *The Hidden Dimension,* Arthur Koestler's *The Ghost in the Machine,* George B. Leonard's *Education and Ecstasy,* Konrad Lorenz's *On Aggression,* Desmond Morris's *The Naked Ape,* Hans Palmstierna's *Plundring, Svält, Förgiftning,* Gordon Rattray Taylor's *The Biological Time Bomb,* Fredric Wertham's *A Sign for Cain,* and R. Buckminster Fuller's *Operating Manual for Spaceship Earth*—all appearing within the last few years, have redefined man's relation to man and to his environment in new and startling ways. The interdependence of various disciplines can best be illustrated by a story Bucky Fuller likes to tell:

> In the last decade, two important papers were presented to learned societies, one on anthropology and the other on biology. And both these researchers were working completely independently. But it happened by chance that I saw both papers. The biological one was looking into all the biological species that have become extinct. The anthropological one was looking into all the human tribes that had become extinct. Both researchers were trying to find a commonality of causes for extinction. Both of them found the same cause independently— extinction is a consequence of over-specialization. As you get more and more over-specialized, you inbreed specialization. It's organic. As you do, you outbreed general adaptability.
>
> So here we have the warning that specialization is a way to extinction, and our whole society is thus organized . . .

Man is a generalist. It is his extensions (tools and environments) which are designed, that help him to achieve specialization. But by misdesigning these tools or environments, we often achieve a closed feedback loop, and the tools and environments in turn affect men and groups in a way that turns them into permanent specialists themselves. The potential of any device, tool, or environment can be studied before it is structured or manufactured. In fact, computers now give us the ability to build mathematical models of processes, interactions, and systems and to study them beforehand. The recent strides made in the social sciences are providing greater insights into that which is societally valuable.

For thousands of years philosophers, artists, and designers have argued about the "need for beauty," or aesthetic values, in the things we use and live with. One only has to look out the window, or for that matter, back into one's own room, to see where this preoccupation with the look-of-things has led us: *The world is ugly, but it doesn't work well either!* In a world brought nearly to its knees by abject want, a preoccupation with "making things pretty" is a crime against humanity. But (as we have seen in our function complex in Chapter One) man needs structures and devices that are enriched beyond the severely utilitarian.

Delight, balance, and that pleasing harmony of proportions that we project outward into the world and are told to regard as the Eidetic Image, are psychological necessities for us. And not only a creature as sophisticated as man, but lower species as well, seem to need this aesthetic and associational enrichment. Here is a description of this mechanism among birds, as quoted by a leading philosopher-naturalist:

> Everyone knows that most birds build houses, and very efficiently, too. Although not usually artistic, their nests are careful and often ingenious. The tailorbird puts nesting material inside a large leaf, then sews up the edges in a curve so that the leaf cannot unroll. The South American ovenbird, which weighs less than three ounces, makes a nest weighing between seven and nine *pounds*, out of a hollow ball of earth fixed to a branch. In Australia the rock warbler makes a long hanging nest and attaches it to the roof of a cave by spiders' webs; the reaction of the spiders is not described. On the Malay Peninsula the megapodes build artificial incubators: piles of vegetation mixed with sand, which gradually decay and keep the eggs warm. The birds themselves are not as big as ordinary fowl, but the nests can be eight feet high and twenty-four feet across, composed of five tons of material scratched together from a radius of several hundred yards. The house martin builds a neat little house of clay with a front door. A simple nest, like that

of the redstart, means six hundred separate flights for material.

Some birds, however, go further, and build simply for aesthetic effect. These are the bowerbirds of Australia and New Guinea. They are perching birds, between eight and fifteen inches long, which look rather like our own woodpeckers, but are more handsomely costumed. Their specialty is unique. The males make clearings in the forest, and at their edges build elaborate arbors of grass and leaves. On the clearings and in the arbors they set out decorations, carefully chosen and grouped: the heads of blue flowers, shells or brilliant objects such as pieces of glass, cartridge cases, and even glass eyes (though these are harder to come by). The scientist who has studied them most closely, A. J. Marshall, shows pretty clearly that this is simply a variation of sexual display intended to attract the little female, to mark off each particular male's own territory, and to allow him a proper stage on which to display his plumage and his masterful poses. And yet Marshall is bound to admit that the birds seem to enjoy their arbors; that their building goes beyond mere functionalism; and that they display very marked discrimination, which can only be called aesthetic choice, in decorating their bowers. An American collector in New Guinea was making his way through the jungle without thinking of bowerbirds or ever having seen one of their structures, when he suddenly came on a place where the undergrowth had been neatly cleared away from an area some four feet square, and a hut-shaped bower had been built beside it, about three feet tall and five feet broad, with an opening a foot high. "This curious structure fronted on the cleared area. The impression of a front lawn was heightened by several beds of flowers or fruit. Just under the door there was a neat bed of yellow fruit. Further out on the lawn there was a bed of blue fruit. Off to one side there were ten freshly picked flowers." Later this explorer saw the architect returning to its bower. The first thing it did was to notice a match that had been carelessly thrown into the middle of its clearing. It hopped over, picked up the match and, with a toss of its head, threw it out of the clearing. So the explorer collected some pink and yellow flowers and one red orchid, and put these in the clearing. Soon the bird came back and flew straight to the new flowers. It took all the yellow ones and threw them away. Then, after some hesitation, it removed the pink ones. Finally it picked up the orchid, decided not to throw it away with the rest, and spent some time carrying it from one pile to another of its own decorations, until it found one where it would fit in with advantage.

Does that sound incredible? There are other facts about the bowerbirds which far surpass it. After one male has completed his arbor he must guard it, for if he flies off in search of food, a rival male will wreck his bower and steal his decorations. Some species not only decorate their bowers but paint them, with coloured fruit pulp, charcoal powder from burnt logs, and (near homes in Australia) stolen bluing. If a flower in the display fades, it is removed at once; and if a human being interferes, the result of this interference is rectified. One observer took some moss out of a bower and hung it some distance away in the forest. Time and again a radiantly coloured male bird angrily put the moss back. And then the same observer conducted an experiment which I can only call brutal. He set fire to three of the bowers. In each case, a male bird flew out of the trees and perched close by the burning arbor, "his beautiful head bowed and wings dropped, as though sorrowing over a funeral pyre." O Science, what crimes are committed in thy name!*

More controlled experiments have been carried on to prove the importance of aesthetically enriched environments. Recent work done by Professor David Krech at the University of California at Berkeley has provided a multitude of insights. Krech assembled two groups of laboratory rats. One group was brought up in a "deprived" environment, similar to conditions existing for human beings in American slums

* The difficulty with writing a book in many parts of the world is that source material sometimes disappears. The lengthy and charming study of bowerbirds quoted above just had to be included. But the book from which it came is gone irretrievably. Whether it gently floated from Viken towards Denmark or was left behind after a shadow-puppet play in Ubud (Bali) matters little; the fact remains that I would like to acknowledge and request permission to quote it, but knowing neither author nor title, this is somewhat difficult.

and ghettos. The rats were crowded, sanitary conditions were absent or nearly absent, food was uninteresting and meager. Their cages were in perpetual gloom, and shrill, unstructured sounds of a decibel level far too high interrupted them during both waking and sleeping. The second group of animals were brought up in an "enriched" environment. Here colors, textures, and materials had been chosen with great care. Food and water were plentiful, vitamin-enriched, and plenty of space was set aside for family grouping. Soft and pleasant music was piped into their habitat, and changing lights and colors further enhanced the environment.

The result of this experiment showed that members of the second "enriched" group had greater learning capacity, a faster mental development, greater flexibility and adaptability to new stimuli, and far better memories. They also maintained their greater mental capacity into old age. In fact, even their offspring, brought up under normal laboratory conditions, maintained a sizeable lead over the offspring of the "deprived" rats who were also brought up in standard ways. Dissection showed that the size and weight of the cerebral cortex of the enriched rats (the part of the brain responsible for a rich flow of association) was larger, heavier, and more convoluted.

When this experiment was repeated, retaining the differences in environment, but feeding identical amounts of water and identical food to both groups of rats, results were almost identical to the first experiment. In both cases the environment-enriched rats developed a high concentration of an important brain enzyme responsible for the growth of brain tissue. The experiment showed conclusively that the environment alone and its relationship to the rats can change the basic brain chemistry.

Although these experiments could not be done on humans, ghettos, slums, most child-care centers, kindergartens, nursery schools, and, in fact, most schools do recapitulate the environment provided for the "deprived" rats. Most parents (considering schools to be only permanent baby-sitting agencies) never ask whether the teachers are robbing their children of potential brain tissue!

In fact, the rats' deprived environment can be said to exist (for human beings) over 90 per cent of the world. During the last 25 years or so, man-made environments have begun taking on the characteristics of a natural ecology: they are interlocking, user-responsive, and self-regenerating. All of humanity is fed into this new ecology, with little forethought as to how a biological mechanism responds to being ripped out of one habitat and forcefully compelled to exist in another. But we have only to look at our zoos. . . .

Apologists for both schools-as-they-are and for slums (and they *do* tend to be the same people) often explain that life is grim and earnest, that existence is a continuous battle where the strong reap victory, and that the young are merely being taught to be tough in order to survive more easily in a tough world. Certainly we have managed to make life grim and earnest, aided by 2,000 years of Judeo-Christian moralizing and sermonizing. But with the advent of more leisure and the prospect of abundance for all, life will surely take on the qualities of joyousness, awareness, uniqueness, self-actualization, communication, empathy, non-conditional love, and transcendental ecstasy. The concept that the strong will perpetually triumph over the weak (". . . a boot stamping on your head, forever . . .") is partially based on a perversion of Darwin's *Origin of Species* theory, "survival of the fittest," as consciously misinterpreted by the rising capitalist class in late nineteenth-century England and America. Partially it arises from the concept that there "is not enough to go around," a historical fact until recently. But the fact of the matter is that today there is more than enough to go around for everyone if only it is properly planned, distributed, and consumed. There is a second fallacy in the concept of the school as a toughening-up ground for the hazards of life.

According to Dr. M. W. Sullivan, quoted in George Leonard's *Education and Ecstasy*, during World War II members of the United States Marine Corps fighting in the South Pacific were exposed to some of the most insufferable conditions in history. Climate, vegetation, and wildlife made life nearly unbearable; the added hazards of battle and disease were incredible. A study showed that the men coming from deprived environments, in other words, those who had been "toughened up for life," were the first to crack up. The Marines whose background had been both an enriched and a more tranquil one more easily withstood the ravages of environment and enemy action. The same experience also has been documented by Dr. Bruno Bettelheim for the inmates of Nazi extermination camps, and held true of captured American soldiers during the Korean War (cf.: *In Every War But One* by Eugene Kincaid, published in 1959 by Norton).

In a dramatically changing world society that is (tremblingly) afraid of change and that educates its young into ever-narrowing areas of specialization, the integrated, comprehensive, anticipatory designer is a dedicated synthesist. Much of the hope lies in the fact that a society grown too large and complex to understand itself or to respond to new events is often unaware of the changes taking place within it. Thus, while much publicity has been given to the fact that more than half of all the people alive today are 25 years old or younger, that by 1986 considerably more than one third of all the people will be less than 15 years old, that even today China has more children under 10 than the total all-age population of Russia and the United States combined, the world has made no relevant responses to these facts. Today there are more college students in the United States than there are farmers. Yet the overly generous subsidies provided for American farmers (at a time when agricultural workers accounted for 98 per cent of the population, rather than today's 7 per cent) are still enforced. The student population is treated to tear gas and clubbings by the police. Buckminster Fuller observes: "Each child today is born in the presence of less mis-information." The accelerated up-grading of so sizeable a part of the population in our schools and universities will inescapably affect all our systems.

Much is done by the power structure, both inside the schools and out, to keep young people from either realizing their power or fulfilling their potential. One answer is war.

"Every 20 years or so we scrap a generation by violent and expensive means, and very soon it is the expense and not the scrapping that bothers us" (Michael Innes). And in the universities we teach narrow, specialized vocational skills (with the emphasis on "earning a living") while paying lip service to "educating the whole man" (in order to supplement the skills taught and turn the students into competent consumers).

The fact is that nearly all of us are so victimized by the propaganda of the profit system that we are no longer able to think straight. During the summer of 1969 when the Swedish government acquired a 10 per cent slice of the Swedish pharmaceutical industry, a leading paper in Stockholm pushed the panic button, saying that if Sweden's entire drug industry were to be socialized, why then "they would only produce what is needed." (!) While ridiculous the point is well taken. For in industrial circles today, most major research concerns itself not with producing for discovered needs, but rather with propagandizing people into desiring what has been produced. *If industry in all countries were to "produce only what is needed," the future would look bright indeed.*

However, members of the industrial design profession continue to support and, in fact, lead the profit-seeking system. David Chapman is the owner and director of one of the largest design firms in the United States. He is a Member of the Board of the Industrial Designers Society of America and has been elected a fellow of both the Royal Society of Arts in England and the International Institute of Arts and Letters in Lindau, Germany. Here is what he has to say* about what he considers to be true market needs:

> The gift market is another enormous area. In 1966, exclusive of Christmas presents, 90 million people received 107 million gifts. Over 40 per cent of table appliances are gifts, even though nobody packs or designs them as

---

* In *Design Seminar*, a report published by the American Iron and Steel Institute, pp. 4–5.

gifts. They're designed with some stubborn suspicion that they're meant to *work*. Well, they are—but who *needs* a blender? (Italics by Chapman)

He continues, somewhat dejected, over a market that does not, alas, yet exist:

There are 35 million pets in the United States. The owners of those pets spend $300 million a year on pet food, but *only $35 million each year on pet "things." No one has offered the owner a thing to buy for Rover.* It is probably possible to get mink collars at Neiman-Marcus, *but no such merchandise crosses America.*

Mr. Chapman also talks about the food needs of the United States. After explaining that "the kitchen is as dead as a dodo" and that "the kitchen business—just as the buggy whip—is on its way out," he says that we shall all eat TV dinners. However, he adds reassuringly: "Mamma may give the food a pinch of oregano or shot of sherry for womanly, psychological reasons."

"Designers must learn a lot more about the effect of social factors on products and markets," he continues, and adds, "there are 75 million Americans over 45—25 million of them over 65. *They have dentures, stomach trouble and things like that. It's a whole new market* and they have lots of money to spend on the things they want" (my italics). Having thus explored the problems of nutrition, the elderly, the sick, and the needy (!), Mr. Chapman triumphantly concludes:

On a new car, for instance, the list price was recently $2,500, but with the extras the car cost $4,200. Who *needs* whitewalls? They don't last longer, *they look cuter.* It is possible to mis-recognize the kind of animal we are all dealing with. Basically, it is a creature seeking total indulgence.

When Mr. Chapman uses words like "animal" and "creature," he is talking about you and me: consumers, clients, his public.

Historical note: Because of many outraged letters, telephone calls, and even one telegram I received in the past, accusing me of inventing both Mr. Chapman and the quotes given above, I should like to confirm that David Chapman really exists, and he really is a fellow of the Industrial De signers Society of America, an honorary member of man international design societies, and an esteemed spokesman of the American Design Establishment. Moreover, Mr Chapman was not being sarcastic in any of the above com ments; in fact, he went to the trouble of having them printed up in a pamphlet (called "Design Seminar") an had his office mail out hundreds of copies to fellow designer and students.

Actually, his remarks are, if anything, a great deal more moderate than those of others in his field. More extreme viewpoints dominate the field, the designers' societies, pro fessional meetings, and, what is most disturbing, most de sign schools in the United States today. Unblushingly, in dustrial design in America has elected to serve as pimp for big business interests.

Ironically, most of the real "plums" or "glamour jobs" the majority of industrial design students in the United States are educated to deal with and delighted to nab just happen to be with blue-chip American firms whose policies and practices are far from progressive when it comes to respect ing the public interest generally and people's need for low cost, ecologically wise,* and aesthetically pleasing products In fact, many American corporate giants have been involved in litigation with the government on charges of either anti trust violations or product liability suits. However, even when anyone manages to win convictions against such com panies in court, punishment often seems ridiculously light In other words, in teaching industrial design as we do, we prepare young people to aid and abet people who fail to measure up to even the prevailing minimal standards that our judicial agencies so feebly enforce.

Here is one example: in the spring of 1970 the big 3 automobile firms were charged before the Supreme Court.

* On June 30, 1971, when firms that pollute were to file state ments with the Federal Government, only 50 out of an esti mated 80,000 had bothered to do so.

They were accused of having conspired for 17 years to keep anti-pollution devices off the market. The three firms admitted this. They begged the courts, however, not to prosecute in exchange for their promise that they would try harder (!), presumably during the next 17 years.

One gratifying fact is that many young people studying design today are unwilling to go on being fed the pap that the schools dispense so readily. As Bill Blau suggested in his piece in *Fortune* a few years ago, the role of this old-fashioned design is slowly coming to an end. If we list a few of the new generation of products to be expected within the next ten years at most, and if we furthermore restrict this listing to products serving *only the Western world,* we will find:

Hovercraft
Monorail systems
Ultra-compact electric cars
Personal, battery-driven mobility devices, that can easily
    be hand-carried
Mass-produced multiple-use buildings
Automated traffic
Computerized medical diagnostic devices
Television-phones
Computer-access consoles in the home
Education through television and teaching machines
De-polluted manufacturing systems
Wide use of bio-degradable materials

The effect of these new products would be to leave us with completely obsolescent roads, automobile factories, schools, universities, housing, factories, hospitals, newspapers, magazine and book publishers, stores, farms, railroad systems, etc. It is not difficult to see why big business is afraid of changes that may phase out its plants and products *as we now know them.*

As factories and industrial combines grow in size, complexity, and investment capital, their opposition to innovation grows. Changes in the system, replacements of the system itself or parts of it become more costly to contemplate and more difficult to institute. Directions of change therefore cannot be expected to be initiated by big business or the military-industrial complex (or the tame, captive designers working for them) but will be initiated by the design team.

To do the most effective job possible, a great deal of research will be needed. A great many questions (most of them trans-national in character) need to be asked. All of these are rather big questions indeed:

What is an ideal human social system? (This will mean an in-depth study of such diverse social organizations as American Plains Indians, the Mundugumor of the Lower Sepik River basin; the priest-cultures of the Inca, Maya, Toltec, and Aztec; the Pueblo cultures of the Hopi; the social structuring surrounding the priest-goddess in Crete; the mountain-dwelling Arapesh; child care in Periclean Greece; Samoa of the late nineteenth century, Nazi Germany, and modern-day Sweden; hunting customs among Australian aborigines, Bantu, and Eskimo; the place of authority and decision-making in China, imperial Rome, slums, and ghettos, and the Loyalist Regime in Spain; delegation of authority in armies, the Catholic Church, modern industrial networks; etc., etc.)

What are optimal conditions for human society on earth? (An inquiry into living patterns, sexual mores, world mobility, codes of behavior, primitive and sophisticated religions and philosophies, and much more will be needed here.)

What are the parameters of the global ecological and ethological system? (Here new insights from such diverse disciplines as meteorology, climatology, physics, chemistry, geology, Von Neumann's Game Theory, cybernetics, oceanography, biology, and all the behavioral sciences will be urgently required; as well as ways of establishing links between these disciplines.)

What are the limits of our resources? (Studies comparable to those carried on by the World Resources Inventory Center at Southern Illinois University will have to be brought into continuous contact with changing technologies and new discoveries.)

What are the human limits?

*What are the basic housekeeping rules for human life on
the planet earth?* (Or, in Bucky Fuller's phrase: *An Operating Manual for Spaceship Earth.*)

And, finally, what don't we know?

There are very few answers to any parts of these questions
as yet. But the first beginnings have been made in creating
tools that may help to begin giving us answers. The International Geophysical Year and the International Years of
the Quiet Sun and the International Upper Mantle Project
were all recent scientific data-gathering attempts of a
trans-national character. Agencies already exist. UNESCO,
UNICEF, the World Health Organization, the International
Labour Organisation, the Scientific Committee of Water Research, the International Council of Scientific Unions, the
Intergovernmental Oceanographic Commission, the International Committee of Manpower Resources are just a few
of some of the organizations now in existence who gather,
store, and retrieve data of global importance.

There is no question but that an International Council of
Anticipatory Comprehensive Design should be established
at the earliest moment. It might well be partially funded by
and work with UNESCO.

But doing the gigantic research task is only one third of
the job that needs to be done to come to grips with the needs
of the world.

The second is the immediate pre-empting of presently
wasted design efforts, and the redirection of these efforts
towards short-range practical design needs. One way of
achieving this at once has been suggested in Chapter Four
as *kymmenykset*. It suggests that designers and design
offices immediately begin turning at least one tenth of their
talents and working time towards the solving of those social
problems that may yield to design solutions. Furthermore,
it means (as suggested in Chapter Ten) that designers refuse to participate in work that is biologically or socially
destructive (whether directly or by implication is of no
importance).

Just this would be a gigantic step towards the common
good. We have marveled together, in an earlier chapter, that

merely by eliminating the rotting of food and by stopping
the destruction of food by vermin, the total protein intake
of billions now suffering would be raised from starvation to
nutritionally acceptable levels. The same can be done in
design. Merely by eliminating the social and moral irrespon-
sibility now prevalent in what I'm tempted to call *all* design
offices and schools, the needs of the neglected half of the
world could be met.

Finally, and as our third point, completely new directions
must be explored in the education of young designers. While
this topic has been given an entire chapter to itself, some
further observations are in order.

The unchecked growth of schools, colleges, and universities has created an environment that is harmful to innovation or, for that matter, education. The problem of size alone
(the university at which I used to teach has 27,000 students
and there are universities more than 3 times as large) work
against education. It tends to make students feel like cogs
in a machine, reduces them to numbers, and alienates them.
This fragments their efforts, and a true learning situation
cannot arise. At the other end of the scale there are private
schools which are considered "small" with between 500 to
3,000 students. These institutions substitute exclusiveness
and the atmosphere of a country club for the giantism of
state universities. The third type of school is usually a highly
specialized one, dealing with the specific problems of the
arts, crafts, or what-have-you. These schools suffer from a
lack of broad general resources and subject matter and tend
to perpetuate the exclusivity of artists-craftsmen and the
formation of little cliques. The fourth possibility, as set up
in London in July of 1969, is a university open to all, where
courses are taken via correspondence, radio, and television.
This last model effectively removes all interaction between
students, or students and teachers.

In all likelihood there are reasons and needs in our society that can be used to justify all four of these methods
of teaching. But young people are forced to make a choice
between size and exclusiveness.

Alternate ways of learning and interacting are already

eing found in many places. The Esalen Institute at Big
ur, California, conducts a peripatetic seminar in the be-
avioral sciences, psychotherapy, and self-awareness.
ranches of Esalen have been set up at San Jose, at Stanford,
nd in San Francisco. Similar institutes exists in over 250
ities. The growth of the Human Potential Movement is
ne of the more startling phenomena of the last few years.
At least one school, the School of Design, California Insti-
ute of the Arts, is attempting to build the behavioral sci-
nces and social design into its regular design curriculum.

In today's renaissance of crafts, weaving, silversmithing,
lassblowing, ceramics, and sculpture are both practiced
nd taught in small centers that are directed primarily to-
vards the summer vacation "trade." Such centers exist in
Maine, California, New Mexico, Michigan, Wisconsin, and
North Carolina, and new ones are springing up all the
ime. The Penland School of Crafts in Penland, North
Carolina, may well be the most successful one of these.
Through its summer tuitions, it supports a group of "resi-
dent craftsmen" throughout the other 9 months of the year.
Penland is dedicated to a free mix of professional craftsmen, craft teachers, college students, retired couples, little
old ladies in tennis shoes, and world-renowned designers. It
also acts as a germinal force in re-establishing a "cottage
industry" based on the crafts in the nearly inaccessible small
arms in southern Appalachia.

Many of the communes are craft-oriented; some of them
manage to support themselves through what they make.

*New Schools Exchange Newsletter, The Whole Earth Cata-
og,* the *Canadian Whole Earth Almanac, Mother Earth
News, Green Revolution, Modern Utopian,* and *The Alternate
Society News* are a few of the channels of communication
and information that are being built to by-pass existing but
outmoded forms.

Frank Lloyd Wright tried to create a milieu that would be
conducive to the study of architecture and planning at
Taliesin and Taliesin West. Unfortunately this experiment,
lasting some thirty years, was too strongly overshadowed
by Mr. Wright's own powerful personality. With this (architectural) exception, the study, research, and practice of

design and planning as socially and morally responsible ac
tivities have not been attempted so far.

It seems crucial that such an experimental design milieu
be established somewhere in the world at once. I envision i
less as a school than as a working environment. Here, youn
people would "learn" through working on real design prob
lems rather than artificially constructed exercises. Such a
working environment would, of necessity, be small in size
at no time accepting more than 30 "students" at most. Part
although a minor part, of its function would be to serve a
a prototype for similar environmental design workshops t
be set up as an interacting global network. Ultimately
students might then have the choice between one schoo
with 30,000 students versus 1,000 environments of 3
students each.

The young people coming to this first, prototypal schoo
would come freely from all parts of the world. They woul
stay for a year or longer and participate in the simultaneou
learning and practicing of integrated design. These youn
men and women would be of varied backgrounds, differin
age groups, with study and work experiences in many differ
ent fields. At all times they would operate as a multi-disci
plinary design team. Their work would be socially relevan
and always "real." By this I mean that, rather than settin
to work on theoretical problems chosen only for their simi
larities to problems dealt with in professional design office
(as is done in all schools), members of the team would di
rect their attention to the actual needs of society. In othe
words, all the work carried on in this milieu would be an
ticipatory.

Such an environment would satisfy a major social nee
not filled today: the creation of a body of designers traine
in the skills that the future will demand of them. Just a
astronauts and cosmonauts are taught skills that may b
demanded of them months or years hence on the moon o
Mars, the design team too will have to prepare itself for th
social challenges of integrated comprehensive design tha
the future will bring. The solutions of design problems wil
be turned over to concerned individuals, social groups, gov
ernments, or trans-national organizations.

As this entire concept of an experimental design environment is thought of as non-profit-earning, any money "earned" through solving these real problems would be directly returned to the work group as tools, machinery, devices, structures, and land. We only have to examine learning situations which people find rewarding, "fun," and in which they learn optimally, to see why the small size of this group is important.

Earlier in this book I discussed learning to drive a car. This skill is taught on a one-to-one, teacher-student ratio. It is further reinforced by the equipment used (the car) and the environment. Other, similar, valuable learning situations are ski schools and swimming schools. Here again the emphasis is on a small teacher-student ratio, a mutually interactive and mutually reinforcing group, and the action of this group within the environment. Most importantly, perhaps, the "teacher" possesses and practices the identical skills which the "students" are learning. He is never a remote professor, tied up within the ivory tower of his own research (as is the case in the universities). Nor is he a "teaching assistant" or graduate student so busy with his own studies that he can give only scant attention to his students.

There is no question that teachers (especially in design) must be constantly involved in its practice. But only a system such as the one proposed here will eliminate the false divorce between practice and teaching.

All members of this team would live and work communally. Their existence would be eased through the whole concept of "communal sharing": that is, consuming more, but owning less. A representative group of 30 present-day university students will serve as one small example: they own, on the average, 26 automobiles, 31 radios, and 15 high-fidelity systems. Without belaboring the obvious, such a capital investment in transient consumer goods would eliminate itself. While expediency would demand the starting of such a "school" in a series of old buildings, a farm, or the like; the eventual buildings would be the responsibility of the team. Temporary domes, information-input cubes (à la Ken Isaacs), and the constructing of more permanent working rooms, sleeping spaces, and social spaces would provide tean members with valuable experiences in a living-working en vironment—one that is constantly changing, constantly be ing questioned and experimentally restructured througl their own thinking and their own labor.

Their "curriculum" would be a loosely woven mesh o those activities and skills needed for creative problem-solv ing. There could be no separation between their "work" anc their leisure-time activities. The newest methods of data processing, film-making, etc., would be available to the team Such a center of design research and planning would hav to be able to offer its hospitality freely to specialists from many disciplines. Such concerned workers could then be drawn into the working and living experiences of the tean for a few days, weeks, or even a year. Because of the experi mental nature of the various structures making up the en vironment, such a center would best be located in the country, but close enough to major urban centers to partici pate in studies, internship work, and experiences in the city environment. What is studied, and how, would evolve or ganically out of the needs of society. There could never be a static "plan of study."

There is no question but that within two or three years some members would leave, their minds full of ideas for a better way of running such an environment. This is unavoidable and would bring about dynamic changes. For it is my belief that if such a center were to be established, soon similar centers would "spin-off." These new centers would be able to address themselves to local and regional problems around the world. They would form the first links in a network of such environments. At each center, young people would be encouraged to travel widely. Such travel could well include a few months' or years' stay and participation in the work a another center. Two things are proposed here: the establish ment of a learning-working environment for thirty young people; and, optimally, a new life style for the peoples of the world.

In the preceding chapter I have explored the dynam ics of the integrated designer's methods of problem-solving and diagrammed them. By now, it will be obvious that I have written this entire book according to this same diagram (as shown below). It has been derived from the in-put of many flow charts. (If it lacks a smooth, linear sequence, it may be put down to that.) The task at hand has been to present you, the reader, with a collection of jig-saw puzzle pieces, which I urge you to put together in whatever pattern seems most relevant. There is no other way of presenting the simultaneity of events.

**DIAGRAM OF ONE CYCLE OF
INTERLOCKING DESIGN "EVENTS".**

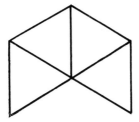

Books like this are expected to end with a dazzling view of the future, and ordinarily this would be the place to speak about vast cities under the ocean, colonies on Mars and Proxima Centauri, machines that will provide us with an everlasting cornucopia of electronic gadgets. But that would be insane.

Design, if it is to be ecologically responsible and socially responsive, must be revolutionary and radical (going back to the roots) in the truest sense. It must dedicate itself to nature's "principle of least effort," in other words, minimum inventory for maximum diversity (to use Peter Pearce's good phase) or, doing the most with the least. That means consuming less, using things longer, recycling materials, and probably not wasting paper printing books such as this.

The insights, the broad, non-specialized, interactive overview of a team (heritage of early man, the hunter) which the designer can bring to the world must now be combined with a sense of responsibility. In many areas designers must learn how to de-design. In this way we may yet have "Survival Through Design."

Bibliography

Bibliography

THIS BIBLIOGRAPHY is startlingly long; it runs to almost five hundred titles. Having just written a book on design as a multi-disciplinary approach, I have tried to make
the bibliography multi-disciplinary as well. Consequently,
books dealing with ecology, ethology, economics, biology,
planning, psychology, literature, anthropology, politics, and
the behavioral sciences are listed together with books on the
future, the environment, popular culture, and design.

These books are not a "suggested reading list" because
they reflect entirely too many personal preferences of my
own. (Nor does this bibliography attempt to list all the books
I have read that have led me to form my convictions. Obviously my own reading list has been considerably larger
than suggested by this listing.) Much has had to be left out.
This might be the place to mention the influence of Tolkien's
Lord of the Rings trilogy, Robert Heinlein's *Stranger in a
Strange Land,* Göran Palm's *As Others See Us,* and almost
everything my friend Henry Miller has written.

The function of this bibliography is to suggest half a
thousand books, none of which would make a bad beginning
for a designer or a student wishing to read into other areas.

Not many books about design are represented here. But
listing all the books on design would mean listing some of
the worst ones, repeating bibliographies found elsewhere,
and breaking no new ground in widening the design field to-

wards other disciplines. This would be inexcusable in one of the first books on industrial design to be published in nearly one and a half decades.

Certainly one book published in 1970—*Design in America* (by the Industrial Designers Society of America)—is no more than an expensive coffee table volume, which painstakingly, extravagantly, and unconsciously illustrates all the points that I have made in this book. When a sales-directed, profit-oriented, ecologically disengaged society practices design, books like *Design in America* use beautiful photographs to illustrate the symptoms of our sickness, rather than listing the causes. But it is fatal to mistake the hectic flush of a final coma for the radiant glow of health. . . .

From the linear thinking of the Renaissance (that great setting of the sun, which man mistook for the dawn), when men still thought all their knowledge classifiable, we have inherited our graphs, divisions, classifications, and lists. Typically when we wish to classify areas of knowledge too vast to be so comprehended, we make the crowning mistake: we educate *specialists*.

But as we go towards the year 2000, as we see divisions that the last few generations have painstakingly erected out of the quicksand of their statistician's minds crumble away, we find no need for more such distinct areas but for unity. Not the *specialist* then, but the *synthesist*.

This is the way in which a meaningful organic pattern, unity, synthesis, will grow between you and each book you read. Out of all the battles you have with the author, the enlightenments and insights his book gives you, the mistakes and confusions you will discover in his work, there *will grow a new entity*, and it will be your gain, and your gain alone.

I) STRUCTURE, NATURE, AND DESIGN

Alexander, Christopher. *Notes on the Synthesis of Form.* Cambridge, Mass.; Harvard University Press, 1964.
————. "Systems Generating Systems," in *Systemat.* Inland Steel Co., 1967.
Alexander, R. McNeill. *Animal Mechanics.* Seattle, Wash.: University of Washington Press, 1968.

Architectural Research Laboratory. *Structural Potential of Foam Plastics for Housing in Underdeveloped Areas.* Ann Arbor, Mich.: 1966.

Baer, Steve. *Dome Cookbook* (3rd printing). Corrales, N.M.: Lama Foundation, 1969.

Bager, Bertel. *Nature as Designer.* New York: Reinhold, 1966.

"Bionik," special number of *Urania* magazine, August, 1969. Leipzig, Germany.

Bootzin, D., and Muffley, H. C. (ed.). *Biomechanics.* New York: Plenum Press, 1969.

Borrego, John. *Space Grid Structures.* Cambridge, Mass.: M.I.T. Press, 1968

Boys, C. V. *Soap-Bubbles.* New York: Dover, 1959.

Brand, Stewart (ed.). *The Whole Earth Catalog* (all issues). Menlo Park, Cal.: 1968–70.

Burkhardt, Dietrich; Schleidt, Wolfgang; and Altner, Helmut. *Signals in the Animal World.* London: Allen & Unwin, 1967.

Clark, Sir Kenneth. *The Nude.* Princeton, N.J.: Bollingen Series, Princeton University Press, 1956.

Critchlow, Keith. *Order in Space.* London: Thames & Hudson, 1969.

Cundy, M. Martyn, and Rollett, A. P.: *Mathematical Models.* Oxford, England: Oxford University Press, 1962. (2nd ed.)

Ganich, Rolf. *Konstruktion, Design, Aesthetik.* Esslingen am Neckar, Germany: 1968.

Gerardin, Lucien. *Bionics.* New York: World University Library, 1968.

Grillo, Paul Jacques. *What Is Design?* Chicago: Paul Theobald, 1962.

Hertel, Heinrich. *Structure, Form and Movement: Biology and Engineering.* New York: Van Nostrand-Reinhold, 1966.

Heythum, Antonin. *On Art, Beauty and the Useful.* Stierstadt im Taunus, Germany: Verlag Eremiten-Presse, 1955.

Hoenich, P. K.: *Robot Art.* Haifa, Israel: Technion, 1962.

Holden, Alan, and Singer, Phyllis. *Crystals and Crystal Growing.* New York: Anchor, 1960.

Kare, Morley, and Bernard, E. E. (ed.). *Biological Prototypes and Manmade Systems.* New York: Plenum Press, 1962.

Katavolos, William. *Organics.* Hillversum, Holland: De Jong & Co., 1961.

Negroponte, Nicholas. *The Architecture Machine.* Cambridge, Mass.: M.I.T. Press, 1970.

Oliver, Paul. *Shelter and Society.* New York: Praeger, 1969.

Otto, Frei (ed.). *Tensile Structures*, Volume One, *Pneumatic Structures.* Cambridge, Mass.: M.I.T. Press, 1967.

————. *Tensile Structures*, Volume Two, *Cables, Nets and Membranes*. Cambridge, Mass.: M.I.T. Press, 1969.

Pawlowski, Andrzej. *Fragmenty Prac Naukowo-Badawczych*. Krakau, Poland: 1966.

Pearce, Peter. *Structure in Nature as a Strategy for Design*. Cambridge, Mass.: M.I.T. Press (in press 1971).

Popko, Edward. *Geodesics*. Detroit, Mich.: University of Detroit Press, 1968.

Ritterbush, Philip C. *The Art of Organic Forms*. Washington, D.C.: Smithsonian Press, 1968.

Schillinger, Joseph. *The Mathematical Basis of the Arts*. New York: Philosophical Library, 1948.

Schwenk, Theodor. *Sensitive Chaos: The Creation of Flowing Forms in Water and Air*. London: Rudolf Steiner Press, 1965.

Sinnott, Edmund W. *The Problem of Organic Form*. New Haven, Conn.: Yale University Press, 1963.

Thompson, Sir D'Arcy Wentworth. *On Growth and Form* (2 vols.). Cambridge, England: Cambridge University Press, 1952.

Wedd, Dunkin. *Pattern & Texture*. New York: Studio Books, 1956.

Weyl, Hermann. *Symmetry*. Princeton, N.J.: Princeton University Press, 1952.

Whyte, Lancelot Law. *Accent on Form*. New York: Harper, 1954.
————. *Aspects of Form*. London: Lund Humphries, 1951; New York: Pellegrini & Cudahy, 1951.
————. *The Next Development in Man*. New York: Mentor, 1950.

Zodiac magazine, Volume 19. Milan, Italy: 1969.

2) DESIGN AND THE ENVIRONMENT

Arvill, Robert. *Man and Environment*. Harmondsworth, England: Penguin, 1967.

Boughey, Arthur S. *Ecology of Populations*. New York: Macmillan, 1968.

Calder, Ritchie. *After the Seventh Day*. New York: Mentor, 1967.

Commoner, Barry. *Science and Survival*. New York: Viking-Compass, 1967.

Curtis, Richard, and Hogan, Elizabeth. *Perils of the Peaceful Atom*. New York: Ballantine, 1970.

DeBell, Garrett (ed.). *The Environmental Handbook*. New York: Ballantine, 1970.

Dubos, René. *Man, Medicine, and Environment*. New York: Mentor, 1968.

Ehrlich, Dr. Paul. "Eco-Catastrophe!" *Ramparts* magazine, September 1968; and San Francisco, Cal.: City Lights, 1970.

———. *The Population Bomb.* New York: Ballantine, 1970.

———. *The Eternal Present: The Beginnings of Art* (Vol. I). Princeton, N.J.: Bollingen Series, Princeton University Press, 1962.

———. *The Beginnings of Architecture* (Vol. II). Princeton, N.J.: Bollingen Series, Princeton University Press, 1964.

Giedion, Sigfried. *Mechanization Takes Command.* New York: Oxford University Press, 1948.

———. *Space, Time and Architecture.* Cambridge, Mass.: Harvard University Press, 1949.

Kaprow, Allan. *Assemblage, Environments & Happenings.* New York; Abrams, 1966.

———. *The Beer Can by the Highway.* New York: Doubleday, 1961.

Kouwenhoven, John A. *Made in America.* New York: Doubleday, 1948.

Kuhns, William. *Environmental Man.* New York: Harper & Row, 1969.

Linton, Ron. *Terracide: America's Destruction of Her Living Environment.* Boston: Little, Brown, 1970.

———. *Confessions of a Dilettante.* New York: Harper & Row, 1967.

———. *The Domesticated Americans.* New York: Harper & Row, 1963.

Lynes, Russell. *The Tastemakers.* New York: Harper, 1954; Grosset Universal Library paperbound.

Marine, Gene. *America the Raped: The Engineering Mentality and the Devastation of a Continent.* New York: Simon & Schuster, 1969.

Marx, Wesley. *The Frail Ocean.* New York: Ballantine, 1970.

McHarg, Ian L. *Design with Nature.* New York: Natural History Press, 1969.

Mitchell, John G. (ed.). *Ecotactics.* New York: Pocketbooks, 1970.

———. *The Brown Decades.* New York: Dover, 1955.

———. *The City in History.* New York: Harcourt, Brace, 1961.

———. *The Condition of Man.* New York: Harcourt, Brace, 1944.

———. *The Conduct of Life.* New York: Harcourt, Brace, 1951.

———. *The Culture of Cities.* New York: Harcourt, Brace, 1938.

———. *From the Ground Up.* New York: Harcourt, Brace, 1956.

———. *Sticks and Stones.* New York: Dover, 1955.

Mumford, Lewis. *Technics and Civilization.* New York: Harcourt: Brace, 1934.

Paddock, William and Paul. *Famine 1975!* Boston: Little, Brown, 1967.
Palmstierna, Hans. *Plundring, Svält, Förgiftning.* Örebro, Sweden: Rabén & Sjögren, 1969.
Ramo, Simon. *Cure for Chaos.* New York: David McKay, 1969.
Rienow, Robert, and Train, Leona: *Moment in the Sun.* New York: Ballantine, 1970.
Shepard, Paul. *Man in the Landscape.* New York: Knopf, 1967.
Shepard, Paul, and McKinley, Daniel. *The Subversive Science: Essays Toward an Ecology of Man.* Boston: Houghton Mifflin, 1969.
Shurcliff, William A. *S/S/T and Sonic Boom Handbook.* New York: Ballantine, 1970.
Smithsonian Institution. *The Fitness of Man's Environment.* Washington, D.C.: Smithsonian Press, 1967.
Sommer, Robert. *Personal Space: The Behavioral Basis of Design.* Englewood Cliffs, N.J.: Prentice-Hall, 1969.
Sotamaa, Yrjö (ed.). *Teollisuus, Ympäristö, Tuotesuunnittelu (Industry, Design, Environment)* (4 vols., tri-lingual). Helsinki, Finland: 1969.
Still, Henry. *The Dirty Animal.* New York: Hawthorn, 1967.
Taylor, Gordon Rattray. *The Biological Time Bomb.* New York: Signet, 1968.
United Nations. *Chemical and Bacteriological (Biological) Weapons and the Effects of Their Possible Use.* New York: Ballantine, 1970.
Whiteside, Thomas. *Defoliation.* New York: Ballantine, 1970.

3) DESIGN AND THE FUTURE

Allen, Edward. *Stone Shelters.* Cambridge, Mass.: M.I.T. Press, 1969.
Calder, Nigel. *The Environment Game.* London: Panther, 1968.
——— (ed.). *The World in 1984* (2 vols.). Harmondsworth, England: Penguin, 1965.
Chase, Stuart. *The Most Probable World.* New York: Harper & Row, 1968.
Clarke, Arthur C. *Profiles of the Future.* New York: Bantam, 1960.
Cole, Dandridge M. *Beyond Tomorrow.* Amherst, Wis.: Amherst Press, 1965.
Cook, Peter. *Experimental Architecture.* New York: Universe Books, 1970.
Ellul, Jacques. *The Technological Society.* New York: Vintage, 1967.

Ewald, William R. Jr. *Environment and Change, The Next Fifty Years.*
———. *Environment and Policy, The Next Fifty Years.* all: Bloomington, Ind.: Indiana University Press, 1968.
———. (ed). *Environment for Man, The Next Fifty Years.*
Fuller, R. Buckminster (ed.). *Inventory of World Resources, Human Trends and Needs* (World Science Decade 1965–1975: Phase I, Document 1).
———. *The Design Initiative* (Phase I, Doc. 2).
———. *Comprehensive Thinking* (Phase I, Doc. 3).
——— (ed.). *The Ten Year Program* (Phase I, Doc. 4).
———. *Comprehensive Design Strategy* (Phase II, Doc. 5).
———. *The Ecological Context: Energy and Materials* (Phase II, Doc. 6).
———. *Education Automation.* Carbondale, Ill.: Southern Illinois University Press, 1964.
———. *Ideas and Integrities.* Englewood Cliffs, N.J.: Prentice-Hall, 1963.
———. *Nine Chains to the Moon.* Philadelphia: J. B. Lippincott, 1938.
———. *No More Secondhand God.* Carbondale, Ill.: Southern Illinois University Press, 1963.
———. *Operating Manual for Spaceship Earth.* Carbondale, Ill.: Southern Illinois University Press, 1969.
———. *Untitled Epic Poem on the History of Industrialization.* Highlands, N. C.: Jonathan Williams Press, 1962.
———. *Utopia or Oblivion.* New York: Bantam, 1970.
All the above and future documents: Carbondale, Ill.: World Resources Inventory, Southern Illinois University Press, varying dates.
Hellman, Hal. *Transportation in the World of the Future.* New York: J. B. Lippincott, 1968.
Kahn, Herman, and Wiener, Anthony J. *The Year 2000: Scenarios for the Future.* New York: Macmillan, 1967.
Krampen, Martin (ed.). *Design and Planning.* New York: Hastings House, 1965.
———. *Design and Planning 2.* New York: Hastings House, 1967.
McHale, John. *The Future of the Future.* New York: George Braziller, 1969.
Marek, Kurt W. *Yestermorrow.* New York: Knopf, 1961.
Marks, Robert W. *The Dymaxion World of Buckminster Fuller.* New York: Reinhold, 1960.
Prehoda, Robert W. *Designing the Future.* New York: Chilton, 1967.

Ribeiro, Darcy. *The Civilizational Process*. Washington, D.C.: Smithsonian Press, 1968.
Skinner, B. F. *Walden Two* (fiction). New York: Macmillan, 1948.
Toward the Year 2000: Work in Progress. Being the summer, 1967, issue of *Daedalus* magazine.

4) AGGRESSION, TERRITORIALITY, BIOLOGICAL SYSTEMS, AND DESIGN

Ardrey, Robert. *African Genesis*. New York: Atheneum, 1965.
———. *The Social Contract*. New York: Atheneum, 1970.
———. *The Territorial Imperative*. New York: Atheneum, 1966.
Bates, Marston. *The Forest and the Sea*. New York: Vintage, 1965.
Bleibtreu, John N. *The Parable of the Beast*. New York: Collier, Macmillan, 1969.
Blond, Georges. *The Great Migrations of Animals*. New York: Collier, Macmillan, 1962.
Broadhurst, P. L. *The Science of Animal Behavior*. Harmondsworth, England: Penguin, 1963.
Charter, S. P. R. *For Unto Us a Child is Born: A Human-Ecological Overview of Population Pressures*. San Francisco, Cal.: Applegate, 1968.
———. *Man on Earth*. San Francisco, Cal.: Applegate, 1965.
Darling, F. Fraser. *A Herd of Red Deer*. New York: Anchor, 1964.
Dowdeswell, W. H. *Animal Ecology*. New York: Harper Torchbooks, 1959.
Eiseley, Loren. *The Firmament of Time*. New York: Atheneum, 1966.
———. *The Immense Journey*. New York: Vintage, 1957.
Gray, James. *How Animals Move*. Harmondsworth, England: Penguin, 1959.
Grey, Walter W. *The Living Brain*. Harmondsworth, England: Penguin, 1961.
Hall, Edward T. *The Hidden Dimension*. New York: Doubleday, 1966.
———. *The Silent Language*. New York: Doubleday, 1959.
Koenig, Lilli. *Studies in Animal Behavior*. New York: Apollo Editions, 1967.
Koestler, Arthur. *The Ghost in the Machine*. New York: Macmillan, 1968.
———. *Insight and Outlook*. New York: Macmillan, 1949.
Lévi-Strauss, Claude. *The Raw and the Cooked: Introduction to a Science of Mythology*. New York: Harper & Row, 1969.

Lindauer, Martin. *Binas Språk*. Stockholm: Bonniers, 1964.
Lorenz, Konrad. *Darwin hat recht Gesehen*. Pfullingen, Germany: Guenther Neske, 1965.
———. *Der Vogelflug*. Pfullingen, Germany: Guenther Neske, 1965.
———. *Er redete mit dem Vieh, den Vögeln und den Fischen*. Vienna, Austria: Borotha-Schoeler, 1949.
———. *Man Meets Dog*. London: Methuen, 1955.
———. *On Aggression*. New York: Harcourt, Brace, 1966.
———. *Studies in Animal and Human Behaviour. Volume 1*. Cambridge, Mass.: Harvard University Press, 1970.
———. *Ueber tierisches und menschliches Verhalten* (2 vols.). Munich, Germany: Piper, 1966.
Marais, Eugène. *The Soul of the Ape*. New York: Atheneum, 1969.
Morris, Desmond. *The Biology of Art*. New York: Knopf, 1962.
———. *The Naked Ape*. New York: McGraw-Hill, 1967.
Mumford, Lewis. *The Myth of the Machine: Volume I—Technics and Human Development*. New York: Harcourt, Brace, 1966.
———. *Volume II—The Pentagon of Power*. New York: Harcourt, Brace, 1970.
Storr, Anthony. *Human Aggression*. London: Allan Lane, Penguin Press, 1968.
Taylor, Gordon Rattray. *The Biological Time Bomb*. New York: World, 1968.
Telfer, William, *et al.* (ed.). *The Biology of Organisms*. New York: Wiley, 1965.
———. *The Biology of Populations*. New York: Wiley, 1966.
Tinbergen, Nicolaas. *The Herring Gull's World*. New York: Anchor, 1967.
———. *Social Behavior in Animals*. London: Methuen, 1953; New York: Wiley, 1953.
von Frisch, Karl. *Bees, Their Vision, Chemical Senses and Language*. Ithaca, New York: Great Seal, 1956.
———. *The Dancing Bees*. New York: Harvest, 1953.
———. *The Study of Instinct*. New York: Oxford University Press, 1951.
———. *Man and the Living World*. New York: Harvest, 1963.
Wickler, Wolfgang. *Mimicry in Plants and Animals*. New York: McGraw-Hill, 1968.
Wylie, Philip. *The Magic Animal*. New York: Doubleday, 1968.
Zipf, George K. *Human Behavior and the Principle of Least Effort: An Introduction to Human Ecology*. Boston, Mass.: Addison-Wesley Press, 1949.

5) ERGONOMICS, HUMAN ENGINEERING, AND HUMAN FACTORS DESIGN

Alger, John R. M., and Hays, Carl V. *Creative Synthesis in Design.* New York: Prentice-Hall, 1964.

Asimov, Morris. *Introduction to Design.* New York: Prentice-Hall, 1962.

Anthropometry & Human Engineering. London: Butterworth's, 1955.

Banham, Reyner. *Theory and Design in the First Machine Age.* London: Architectural Press, 1960.

Buhl, Harold R. *Creative Engineering Design.* Ames, Iowa: Iowa State University Press, 1960.

Consumers Union (ed.). *Passenger Car Design and Highway Safety.* Mount Vernon, New York: Consumers Union, 1963.

Glegg, Gordon L. *The Design of Design.* Cambridge, England: Cambridge University Press, 1969.

Goss, Charles Mayo (ed.). *Gray's Anatomy* (27th Edition). Philadelphia: Lea & Febiger, 1959.

Jones, J. Christopher, and Thronley, D. G. *Conference on Design Methods.* Oxford, England: The Pergamon Press, 1963.

McCormick, Ernest Jr. *Human Engineering.* New York: McGraw-Hill, 1957.

Nader, Ralph. *Unsafe at any Speed.* New York: Grossman, 1965.

Schroeder, Francis. *Anatomy for Interior Designers* (2nd Edition). New York: Whitney Publications, 1948.

Starr, Martin Kenneth. *Product Design and Decision Theory.* New York: Prentice-Hall, 1963.

U.S. Navy (ed.). *Handbook of Human Engineering Data (Second Edition) U.S. Navy Office of Naval Research, Special Devices Center,* by NAVEXOS P-643, Report SDC 199-1-2 (NR-783-001. N6onr-199. TOI PDSCDCHE Project 20-6-1) Tufts University, Medford, Mass., n.d.

Woodson, Wesley E. *Human Engineering Guide for Equipment Designers.* Berkeley, Cal.: University of California Press, 1954.

6) GESTALT, PERCEPTION, CREATIVITY, AND RELATED FIELDS

Adorno, T. W. *et al. The Authoritarian Personality.* New York: Harper, 1950.

Allport, Floyd. *Theories of Perception and the Concept of Structure.* New York: Wiley, 1955.

Berne, Dr. Eric. *Games People Play.* New York: Grove Press, 1964.

————. *Principles of Group Treatment*. New York: Grove Press, 1966.

————. *The Structure and Dynamics of Organizations and Groups*. New York: J. B. Lippincott, 1963.

————. *Transactional Analysis in Psychotherapy*. New York: Grove Press, 1961.

Bettelheim, Bruno. *The Empty Fortress: Infantile Autism and the Birth of the Self*. New York: Free Press, 1967.

————. *The Informed Heart: Autonomy in a Mass Age*. New York: Free Press, 1960.

De Bono, Edward. *New Think*. New York: Basic Books, 1968.

Freud, Sigmund. *Beyond the Pleasure Principle* (tr. Strachey). London: Hogarth, 1961; New York: Bantam paperbound.

————. *Moses and Monotheism* (tr. Jones). New York: Knopf, 1939; Vintage paperbound.

————. *On Creativity and the Unconscious*. New York: Torchbooks, n.d.

————. *Totem and Taboo* (tr. Brill). New York: Random House, 1946; Vintage paperbound.

Fromm, Erich. *The Art of Loving*. New York: Harper, 1956.

————. *The Revolution of Hope*. New York: Harper, 1968.

Ghiselin, Brewster (ed.). *The Creative Process*. Berkeley, Cal.: University of California Press, 1952; New York: Mentor paperbound.

Gibson, James J. *The Perception of the Visual World*. Boston: Houghton Mifflin, 1950.

Gordon, William J. J. *Synectics*. New York: Harper, 1961.

Gregory, R. L. *The Intelligent Eye*. New York: McGraw-Hill, 1970.

Grotjahn, Martin. *Beyond Laughter*. New York: McGraw-Hill, 1957.

Gunther, Bernard. *Sense Relaxation*. New York: Collier, 1968.

Jung, C. G. *Archetypes and the Collective Unconscious* (2 vols.). Princeton, N.J.: Bollingen Series, Princeton University Press, 1959.

————. *Psychology of the Unconscious*. London: Kegan Paul, 1922; New York: Dodd, Mead, 1949.

Katz, David. *Gestalt Psychology*. New York: Ronald Press, 1950.

Koehler, Wolfgang. *Gestalt Psychology*. New York: Liveright, 1947, rev. 1970; Mentor paperbound.

Koestler, Arthur. *The Act of Creation*. New York: Macmillan, 1964.

Kofka, K. *Principles of Gestalt Psychology*. London: Kegan Paul, 1935.

Korzybski, Alfred. *The Manhood of Humanity*. Chicago: Library of General Semantics, 1950.

————. *Science and Sanity*. Chicago: Library of General Semantics, 1948.

Kubie, Lawrence S. *The Neurotic Distortion of the Creative Process*. Lawrence, Kans.: University of Kansas Press, 1958; New York, Noonday paperbound.

Leonard, George B. *Education and Ecstasy*. New York: Delacorte Press, 1968.

Lindner, Robert. *Must You Conform?* New York: Rinehart, 1956; Grove Press Black Cat paperbound.

———. *Prescription for Rebellion*. New York: Rinehart, 1952.

Neumann, Erich. *The Archetypal World of Henry Moore*. Princeton, N.J.: Bollingen Series, Princeton University Press, 1959.

Parnes, Sidney, and Harding, H. *A Source Book of Creative Thinking*. New York: Scribner, 1962.

Perls, F. S. *Ego, Hunger and Aggression*. New York: Random House, 1969.

———. *Gestalt Therapy Verbatim*. J. Stephens (ed.). Lafayette, Cal.: Real People Press, 1969.

———. *In and Out of The Garbage Pail*. Lafayette, Cal.: Real People Press, 1969.

Petermann, Bruno. *The Gestalt Theory and the Problem of Configuration*. New York: Harcourt, Brace, 1932.

Rawlins, Ian. *Aesthetics and the Gestalt*. London: Nelson, 1953.

Reich, Wilhelm. *The Cancer Biopathy*. New York: Orgone Institute Press, n.d.

———. *The Function of the Orgasm*. New York: Noonday, 1961.

———. *The Mass Psychology of Fascism*. New York: Orgone Institute Press, 1946.

———. *Selected Writings: An Introduction to Orgonomy*. New York: Noonday, 1961.

———. *The Sexual Revolution*. New York: Noonday, 1962.

Rolf, Dr. Ida P. *Structural Integration*. Santa Monica, Cal.: 1962

Ruesch, Jurgen. *Communication*. New York: Norton, 1951.

———. *Disturbed Communication*. New York: Norton, 1957.

———. *Non-Verbal Communication*. Berkeley, Cal.: University of California Press, 1956

Shanks, Michael. *The Innovators*. Harmondsworth, England: Penguin, 1967.

Smith, Paul. *Creativity*. New York: Hastings House, 1959.

Spence, Lewis. *Myth and Ritual in Dance, Game and Rhyme*. London: Watts, 1947.

Vernon, Magdalen D. *A Further Study of Visual Perception*. Cambridge, England: Cambridge University Press, 1952.

Wertham, Fredric. *Dark Legend*. New York: Paperback Library, 1966.

———. *Seduction of the Innocent*. New York: Macmillan, 1954.

————. *The Show of Violence.* New York: Paperback Library, 1966.
————. *A Sign for Cain: An Exploration of Human Violence.* New York: Macmillan, 1966.
Wiener, Norbert. *Cybernetics.* New York: Wiley, 1948.
————. *The Human Use of Human Beings.* Boston: Houghton Mifflin, 1950; Avon paperbound.

7) POPULAR CULTURE, SOCIAL PRESSURES, AND DESIGN

Adams, Brooks. *The Law of Civilization and Decay.* New York: Vintage, n.d.
Arensberg, Conrad M., and Niehoff, Arthur H. *Introducing Social Change.* Chicago: Aldine, 1964.
Boorstin, Daniel J. *The Image: A Guide to Pseudo-Events in America.* New York: Harper & Row, 1964.
Brightbill, Charles K. *The Challenge of Leisure.* New York: Spectrum, 1960.
Brown, James. A. C. *Techniques of Persuasion.* Harmondsworth, England: Penguin, 1963.
Cassirer, Ernst: *An Essay on Man.* New Haven, Conn.: Yale University Press, 1944.
————. *Language and Myth.* New York: Harper & Brothers, 1946; Dover paperbound.
————. *The Myth of the State.* New Haven, Conn.: Yale University Press, 1946; London: Oxford University Press, 1946.
Goodman, Paul. *Art and Social Nature.* New York: Arts and Science Press, 1946.
————. *Compulsory Mis-education.* New York: Horizon, 1964.
————. *Drawing the Line.* New York: Random House, 1962.
————. *Growing Up Absurd.* New York: Vintage, 1960.
————. *Like a Conquered Province: The Moral Ambiguity of America.* New York: Random House, 1967.
————. *Notes of a Neolithic Conservative.* New York: Random House, 1970.
————. *Utopian Essays and Practical Proposals.* New York: Vintage, 1964.
Gorer, Geoffrey. *Hot Strip Tease.* London: Graywells Press, 1934.
Gurko, Leo. *Heroes, Highbrows and the Popular Mind.* New York: Charter Books, 1962.
Hofstadter, Richard. *Anti-intellectualism in American Life.* New York: Knopf, 1963.
Hofstadter, Richard, and Wallace, Michael. *American Violence.* New York: Knopf, 1970.

Jacobs. Norman (ed.). *Culture for the Millions?* Boston: Beacon,
    1964.
Joad, C. E. M. *Decadence.* London: Faber & Faber, 1948.
Kefauver, Estes. *In a Few Hands: Monopoly Power in America.*
    New York: Pantheon Books, 1965.
Kerr, Walter. *The Decline of Pleasure.* New York: Simon &
    Schuster, 1964.
Kronhausen, Drs. Phyllis and Eberhard. *Erotic Art.* New York:
    Grove Press, 1969.
————. *Erotic Art II.* New York: Grove Press, 1970.
————. *The First International Exhibition of Erotic Art* (Cata-
    logue). Copenhagen, Denmark: Uniprint, 1968.
————. *The Second International Exhibition of Erotic Art*
    (Catalogue). Copenhagen, Denmark: Uniprint, 1969.
Künen, James Simon. *The Strawberry Statement: Notes of a
    College Revolutionary.* New York: Random House, 1969.
Larrabee, Eric, and Meyersohn, Rolf (ed.). *Mass Leisure.* New
    York: Free Press, 1958.
Legman, Gershon. *The Fake Revolt.* New York: The Breaking
    Point Press, 1966.
————. *Love and Death: A Study in Censorship.* New York:
    The Breaking Point Press, 1949.
————. (ed.). *Neurotica: 1948–1951.* New York: Hacker, 1963.
————. *Rationale of the Dirty Joke: An Analysis of Sexual
    Humor.* New York: Grove Press, 1966.
Levy, Mervyn. *The Moons of Paradise: Reflections on the Female
    Breast in Art.* New York: Citadel, 1965.
Macdonald, Dwight. *Masscult and Midcult.* New York: Random
    House, 1961.
Mannheim, Karl. *Ideology and Utopia.* London: K. Paul, Trench,
    Trubner, 1936; New York: Harcourt Brace Harvest edition.
McLuhan, Marshall. *Culture is Our Business.* New York: Mc-
    Graw-Hill, 1970.
————. *The Gutenberg Galaxy.* Toronto, Canada: University of
    Toronto Press, 1962.
————. *The Mechanical Bride.* New York: Vanguard, 1951.
————. *Understanding Media.* New York: McGraw-Hill, 1964.
———— & Carpenter, Edmund. *Explorations in Communication.*
    Boston: Beacon, 1960.
————. & Watson, Wilfred. *From Cliche to Archetype.* New
    York: Viking, 1970.
———— & Fiore, Quentin. *The Medium Is the Massage.* New
    York: Bantam, 1967.
———— & Parker, Harley. *Through the Vanishing Point.* New
    York: Harper & Row, 1968.
———— & Papanek, Victor J. *Verbi-Voco-Visual Explorations.*

New York: Something Else Press, 1967.
———— & Fiore, Quentin. *War and Peace in the Global Village.* New York: Bantam, 1968.
Mehling, Harold. *The Great Time Killer.* New York: World, 1962.
Mesthene, Emmanuel G. *Technological Change.* Cambridge, Mass.: Harvard University Press, 1970.
Molnar, Thomas. *The Decline of the Intellectual.* New York: Meridian, 1961.
Myrdal, Jan, and Kessle, Gun. *Angkor: An Essay on Art and Imperialism,* New York: Pantheon Books, 1970.
O'Brian, Edward J. *The Dance of the Machines.* New York: Macaulay, 1929.
Packard, Vance. *The Hidden Persuaders.* New York: David McKay, 1957; Pocketbooks paperbound.
————. *The Status Seekers.* New York: David McKay, 1959; Pocketbooks paperbound.
————. *The Wastemakers.* New York; David McKay, 1960; Pocketbooks paperbound.
Palm, Göran. *As Others See Us.* Indianapolis, Ind.: Bobbs-Merrill, 1968.
Reich, Charles A. *The Greening of America.* New York: Random House, 1970.
Repo, Satu, (ed.). *This Book is About Schools.* New York: Pantheon Books, 1970.
Riesman, David. *Faces in the Crowd.* New Haven, Conn.: Yale University Press, 1952.
————. *Individualism Reconsidered.* New York: Free Press, 1954.
————. *The Lonely Crowd.* New Haven, Conn.: Yale University Press, 1950; rev. 1961.
Rosenberg, Bernard, and White, David M. *Mass Culture.* New York: Free Press, 1957.
Roszak, Theodore. *The Making of a Counter Culture.* New York: Doubleday, 1969.
Rudofsky, Bernard. *Are Our Clothes Modern?* Chicago: Paul Theobald, 1949.
————. *Behind the Picture Window.* New York: Oxford University Press, 1954.
Ryan, William. *Blaming the Victim.* New York: Pantheon Books, 1971.
Snow, C. P. *The Two Cultures: And a Second Look.* Cambridge, England: Cambridge University Press, 1963.
Thompson, Denys. *Discrimination and Popular Culture.* Harmondsworth, England: Penguin, 1964.
Toffler, Alvin. *The Culture Consumers.* New York: St. Martin's, 1964.

Veblen, Thorstein. *The Theory of the Leisure Class.* New York: Macmillan, 1899.

Wagner, Geoffrey. *Parade of Pleasure: A Study of Popular Iconography in the U.S.A.* London: Derek & Verschoyle, 1954; New York: Library Publishers, 1955.

Walker, Edward L., and Heyns, Roger W. *An Anatomy for Conformity.* New York: Spectrum, 1962.

Warshow, Robert. *The Immediate Experience.* New York: Doubleday, 1963.

Young, Wayland. *Eros Denied: Sex in Western Society.* New York: Grove Press, 1964.

8) DESIGN AND OTHER CULTURES

Belo, Jane. *Traditional Balinese Culture.* New York: Columbia University Press, 1970.

Benrimo, Dorothy. *Camposantos.* Fort Worth, Texas: Amon Carter Museum, 1966.

Bhagwati, Jagdish. *The Economics of Underdeveloped Countries.* New York: McGraw-Hill, 1966.

Carpenter, Edmund. *Eskimo.* Toronto, Canada: University of Toronto Press, 1959.

Covarrubias, Miguel. *Bali.* New York: Knopf, 1940.

————. *Mexico South.* New York: Knopf, 1946.

Cushing, Frank Hamilton. *Zuni Fetishes.* Flagstaff, Arizona: KC Editions, 1966.

Dennis, Wayne. *The Hopi Child.* New York: Science Editions, 1965.

DePoncins, Contran. *Eskimos.* New York: Hastings House, 1949.

Eliade, Mircea. *Shamanism: Archaic Techniques of Ecstasy.* Princeton, N.J.: Bollingen Series, Princeton University Press, 1964.

Herrigel, Eugen. *Zen in the Art of Archery.* New York: Pantheon Books, 1953.

Hokusai. *One Hundred Views of Mount Fuji.* New York: Frederik Publications, 1958.

Iwamiys, Takeji. *Katachi: Japanese Pattern and Design in Wood, Paper and Clay.* New York: Abrams. 1967.

Jenness, Diamond. *The People of the Twilight.* Chicago; University of Chicago Press, Phoenix, 1959.

Kakuzo, Okakura. *The Book of Tea.* Tokyo: Tuttle, 1963.

Kitzo, Harumichi. *Cha-No-Yu.* Tokyo: Shokokusha, 1953.

————. *Formation of Bamboo.* Toyko: Shokokusha, 1958.

————. *Formation of Stone.* Tokyo: Shokokusha, 1958.

Kubler, George. *The Shape of Time*. New Haven, Conn.: Yale University Press paperbound, 1962.

Leppe, Markus. *Vaivaisukot*. Helsinki, Finland: Werner Söderström, 1967.

Liebow, Elliot. *Tally's Corner*. Boston: Little, Brown, 1967.

Linton, Ralph. *The Tree of Culture*. New York: Knopf, 1955; Vintage paperbound.

McPhee, Collin. *A House in Bali*. New York: John Day, 1946.

――――. *Music in Bali*. New Haven, Conn.: Yale University Press, 1966.

Malinowski, Brownislaw. *Magic, Science and Religion*. New York: Anchor, 1954.

――――. *Sex and Repression in Savage Society*. London: Routledge, 1953; New York; Meridian.

Manker, Ernst. *People of Eight Seasons: The Story of the Lapps*. New York: Viking, 1964.

Mead, Margaret: *Coming of Age in Samoa*. New York: Morrow, 1928; Dell paperbound.

――――. *Cultural Patterns and Technological Change*. New York: Mentor, n.d.

――――. *Growing up in New Guinea*. New York: Morrow, 1930; Dell paperbound.

――――. *Male and Female*. New York; Morrow, 1949; Dell paperbound.

――――. *Sex and Temperament*. New York: Morrow, 1935; Dell paperbound.

Michener, James A. *Hokusai Sketchbooks*. Tokyo: Tuttle, 1958.

Mookerjee, Ajit. *Tantra Art*. New Delhi, India: Kumar Gallery, 1967.

Mowat, Farley. *People of the Deer*. New York: Pyramid, 1968.

Oka, Hideyuki: *How to Wrap Five Eggs*. New York: Harper & Row, 1967.

Ortega y Gasset, José. *The Dehumanization of Art* (tr. Weyl). Princeton, N.J.: Princeton University Press, 1948 and revised paperbound edition.

Ortiz, Alfonso. *The Tewa World: Space, Time, Being & Becoming in a Pueblo Society*. Chicago: University of Chicago Press, 1969.

Reichard, Gladys A. *Navaho Religion: A Study of Symbolism*. Princeton, N.J.: Bollingen Series, Princeton University Press, 1950.

Richards, Audrey I. *Hunger and Work in a Savage Tribe*. New York: Meridian, 1964.

Roediger, Virginia More. *Ceremonial Costumes of the Pueblo Indians*. Berkeley, Cal.: University of California Press, 1961.

Rudofsky, Bernard. *The Kimono Mind*. New York: Doubleday, 1965.

Saunders, E. Dale. *Mudra: A Study of Symbolic Gestures in Japanese Buddhist Sculpture*. Princeton, N.J.: Bollingen Series, Princeton University Press, 1960.

Schafer, Edward H. *The Golden Peaches of Samarkand: A Study of T'ang Exotics*. Berkeley, Cal.: University of California Press, 1963.

——. *Tu Wan's Stone Catalogue of Cloudy Forest*. Berkeley, Cal.: University of California Press, 1961.

Spencer, Robert F. *The North Alaskan Eskimo: A Study in Ecology and Society*. Washington, D.C.: Smithsonian Institution Press, 1969.

Spies, Walter, and de Zoete, Beryl. *Dance and Drama in Bali*. London: Faber & Faber, 1938.

Suzuki, Daisetz T. *Zen and Japanese Culture*. Princeton, N.J.: Bollingen Series, Princeton University Press, 1959.

Sze, Mai-Mai, *The Tao of Painting* (2 vols.). Princeton, N.J.: Bollingen Series, Princeton University Press, 1956.

Tange, Kenzo, and Gropius, Walter. *Katsura: Tradition and Creation in Japanese Architecture*. New Haven, Conn.: Yale University Press, 1960.

—— and Kawazoe, Noboru. *Ise: Prototype of Japanese Architecture*. Cambridge, Mass.: M.I.T. Press, 1965.

Watts, Alan R. *Beat Zen, Square Zen and Zen*. San Francisco, Cal.: City Lights, 1959.

——. *The Joyous Cosmology*. New York: Pantheon Books, 1962.

——. *Nature, Man and Woman*, New York: Pantheon Books, 1958; Vintage paperbound.

Wyman, Leland C. (ed.) *Beautyway: A Navaho Ceremonial*. Princeton, N.J.: Bollingen Series, Princeton University Press, 1957.

Yee, Chiang. *The Chinese Eye*. New York: Norton, 1950.

——. *Chinese Calligraphy*. London: Methuen, 1954.

9) PERSONAL STATEMENTS BY DESIGNERS AND OTHERS

Brecht, Bertolt. *Gesammelte Werke*. Frankfurt, Germany: Suhrkamp Verlag, 1967.

Cleaver, Eldridge. *Soul on Ice*. New York: Harper, 1968.

——. *Eldridge Cleaver: Post-Prison Writings and Speeches* (ed. Scheer) New York: Random House, 1969; Vintage, 1969.

Debray, Régis. *Revolution in the Revolution*. New York: Grove Press, 1967.

Fanon, Frantz. *The Wretched of the Earth.* New York: Grove Press, 1963; Harmondsworth, England: Penguin, 1967.
Fischer, Ernst. *The Necessity of Art. A Marxist Approach,* Harmondsworth, England: Pelican, 1964.
Gonzales, Xavier. *Notes About Painting.* New York: World, 1955.
Greenough, Horatio. *Form and Function.* Washington, D.C.: Privately published, 1811.
————. *The Bolivian Diaries of Che Guevara.* New York: Vintage, 1969.
Guevara, Che. *Guerrilla Warfare.* New York: Vintage, 1968.
Kennedy, Robert F. *To Seek a Newer World.* New York: Bantam, 1968.
Koestler, Arthur. *Arrow in the Blue.* New York: Macmillan, 1961.
————. *The Invisible Writing.* Boston: Beacon, 1955.
————. *Scum of the Earth.* London: Hutchinson, 1968.
Laing, R. D. *The Politics of Experience.* New York: Pantheon Books, 1967.
Mailer, Norman. *The Armies of the Night.* New York: Signet, 1968.
————. *Miami and the Siege of Chicago.* New York: Signet, 1968.
Mao Tse-tung. *Collected Writings* (5 vols.). Peking: 1964.
————. *On Art and Literature.* Peking: Foreign Language Press, 1954.
————. *On Contradiction.* Peking: Foreign Language Press.
————. *On the Correct Handling of Contradictions among the People.* Peking: Foreign Language Press.
Marcuse, Herbert. *Das Ende der Utopie.* Berlin, Maikowski, 1967.
————. *One-Dimensional Man.* New York: Bantam, 1964.
Marin, John. *The Collected Letters of John Marin.* New York: Abelard-Schuman, n.d.
Myrdal, Jan. *Confessions of a Disloyal European.* New York: Pantheon Books, 1968.
————. *Report from a Chinese Village.* New York: Pantheon Books, 1965.
————. *Samtida.* Stockholm: Norstedt, 1967.
Papanek, Victor J. "Kymmenen Ympäristöä" (Environments for Discovery), *Ornamo* Magazine (bilingual). Helsinki, Finland: February, 1970.
Richards, M. C. *Centering: In Pottery, Poetry and the Person.* Middletown, Conn.: Wesleyan University Press, 1964.
Saarinen, Eliel. *Search for Form.* New York: Reinhold, 1948.
Safdie, Moshe. *Beyond Habitat.* Cambridge, Mass.: M.I.T. Press, 1970.
St. Exupéry, Antoine de. *Carnets.* Paris: Gallimard, 1953.

———. *Bekenntnis einer Freundschaft*. Düsseldorf, Germany: Karl Rauch, 1955.

———. *Flight to Arras*. London: Heinemann, 1942; New York: Reynal & Hitchcock, 1943.

———. *Frieden Oder Krieg?* Düsseldorf, Germany: Karl Rauch, 1957.

———. *Gebete der Einsamkeit*. Düsseldorf, Germany: Karl Rauch, 1956.

———. *Lettres à l'amie inventée*. Paris: Plon, 1953.

———. *Lettres à sa mère*. Paris: Gallimard, 1955.

———. *Lettres de jeunesse*. Paris: Gallimard, 1953.

———. *The Little Prince*. New York: Reynal & Hitchcock, 1943.

———. *Night Flight*. Harmondsworth, England: Penguin, 1939.

———. *A Sense of Life*. New York: Funk & Wagnalls, 1965.

———. *Wind, Sand and Stars*. New York: Reynal & Hitchcock, 1939.

———. *The Wisdom of the Sands*. New York: Harcourt, Brace, 1952.

Servan-Schreiber, Jean Jacques. *The American Challenge*. London: Hamish Hamilton, 1967. New York: Atheneum, 1968.

Shahn, Ben. *The Shape of Content*. Cambridge, Mass.: Harvard University Press, 1957.

Soleri, Paolo. *Arcology: The City in the Image of Man*. Cambridge, Mass.: M.I.T. Press, 1970.

Sullivan, Louis H. *The Autobiography of an Idea*. Chicago: Peter Smith, 1924.

———. *Kindergarten Chats*. Chicago: Scarab Fraternity, 1934.

Thoreau, Henry David. *Walden* and *Essay on Civil Disobedience*. New York: Mentor; and other editions.

Van Gogh, Vincent. *The Complete Letters of Vincent Van Gogh in Three Volumes*. Greenwich, Conn.: New York Graphic Society, 1959.

Weiss, Peter. *Notizen zum Kultureilen Leben in der Demokratischen Republik Viet Nam*. Frankfurt, Germany: Suhrkamp Verlag, 1968.

Wright, Frank Lloyd. *Autobiography*. New York: Duell, Sloane & Pearce, 1943.

———. *The Living City*. New York: Horizon, 1958.

———. *A Testament*. New York: Horizon, 1957.

Wright, Olgivanna Lloyd. *The Shining Brow*. New York: Horizon, 1958.

Yevtushenko, Yevgeny. *Complete Poems*. New York: Dutton, 1962.

———. *A Precocious Autobiography*. New York: Dutton, 1963.

10) THE BACKGROUND OF DESIGN

Arnheim, Rudolf. *Art and Visual Perception*. Berkeley, Cal.: University of California Press, 1954.

———. *Film as Art*. Berkeley, Cal.: University of California Press, Paperbacks, 1957.

———. *Toward a Psychology of Art*. Berkeley, Cal.: University of California Press, 1967.

Bayer, Herbert and Gropius, Walter. *Bauhaus 1919–1928*. Boston: Branford, 1952.

Berenson, Bernard. *Aesthetics and History*. New York: Pantheon Books, 1948; Anchor paperbound.

Biedermen, Charles. *Art as the Evolution of Visual Knowledge*. Red Wing, Minnesota: 1948.

Boas, Franz. *Primitive Art*. New York: Dover, 1955.

Conrads, Ulrich, and Sperlich, Hans G. *The Architecture of Fantasy*. New York: Praeger, 1962.

Danz, Louis. *Dynamic Dissonance in Nature and the Arts*. New York: Longmans Green, 1952.

———. *It Is Still the Morning* (Novel). New York: Morrow, 1943.

———. *Personal Revolution and Picasso*. New York: Longmans Green, 1941.

———. *The Psychologist Looks at Art*. New York: Longmans Green, 1937.

———. *Zarathustra Jr*. New York: Brentano, 1934.

Dorfles, Gillo. *Il Kitsch: Antologia del cattivo gusto*. Milan, Italy: Gabriele Mazzotta Editore, 1968.

Ehrenzweig, Anton. *The Hidden Order of Art*. Berkeley, Cal.: University of California Press, 1967.

Feldman, Edmund B. (ed.). *Art in American Higher Institutions*. Washington, D.C.: The National Art Education Association, 1970.

Friedmann, Herbert. *The Symbolic Goldfinch: Its History and Significance in European Devotional Art*. Princeton, N.J.: Bollingen Series, Princeton University Press, 1946.

Gamow, George. *One, Two, Three . . . Infinity* (rev.). New York: Viking, 1961; Bantam paperbound.

Gerstner, Karl. *Kalte Kunst?* Basel, Switzerland: Arthur Niggli, 1957.

Gilson, Etienne. *Painting and Reality*. Princeton, N.J.: Bollingen Series, Princeton University Press, 1957.

Gombrich, E. H. *Art and Illusion*. Princeton, N.J.: Bollingen Series, Princeton University Press, 1960.

————. *Meditations on a Hobbyhorse.* London: Faber & Faber, 1964.

Graves, Robert. *The White Goddess.* London: Faber & Faber; New York: Farrar, Straus & Giroux, 1966.

Hatterer, Lawrence J. *The Artist in Society: Problems and Treatment of the Creative Personality.* New York: Grove Press, 1965.

Hauser, Arnold. *The Social History of Art* (4 vols.). London: Routledge, 1951; New York: Vintage, paperbound.

Hogben, Lancelot. *From Cave Painting to Comic Strip.* New York: Chanticleer Press, 1949.

Hon-En Historia (Catalogue). Stockholm: Moderna Museet, 1967.

Huizinga, Johan. *Homo Ludens: A Study of the Play-element in Human Culture.* Boston: Beacon Press, 1950, 1966.

Hultén, K. G. Pontus. *The Machine as Seen at the End of the Mechanical Age.* New York: The Museum of Modern Art, 1968.

Keats, John. *The Insolent Chariots.* New York: Crest Books, n.d.

Klingender, Francis D. *Art and the Industrial Revolution.* London: Noel Carrington, 1947; rev. ed.: London: Evelyn, Adams & Mackay, 1968.

Kracauer, Siegfried. *From Caligari to Hitler.* Princeton, N.J.: Princeton University Press, 1947.

Kranz, Kurt. *Variationen über ein geometrisches Thema.* Munich, Germany: Prestel, 1956.

Langer, Susanne K. *Feeling and Form.* New York: Scribner, 1953.

————. *Philosophy in a New Key.* New York: Scribner, 1942.

————. *Problems of Art.* New York: Scribner, 1957.

Le Corbusier. *The Modulor.* London: Faber & Faber, 1954.

————. *Modulor 2.* London: Faber & Faber, 1958.

Lethaby, W. R. *Architecture, Nature and Magic.* New York: George Braziller, 1956.

Malraux, André. *The Voices of Silence.* New York: Doubleday, 1952.

Maritain, Jacques. *Creative Intuition in Art and Poetry.* Princeton, N.J.: Bollingen Series, Princeton University Press, 1953.

Middleton, Michael. *Group Practice in Design.* New York: George Braziller, 1967.

Moholy-Nagy, Sibyl. *Native Genius in Anonymous Architecture.* New York: Horizon, 1957.

Neumann, Erich. *The Great Mother: An Analysis of the Archetype.* Princeton, N.J.: Bollingen Series, Princeton University Press, 1955.

Neutra, Richard. *Survival through Design*. New York: Oxford University Press, 1954.

Nielsen, Vladimir. *The Cinema as Graphic Art*. New York: Hill & Wang, 1959.

Oakley, Kenneth P. *Man the Tool-maker*. Chicago: University of Chicago Press, 1959.

Ozenfant, Amedee. *Foundations of Modern Art*. New York: Dover, 1952.

————. *Gothic Architecture and Scholasticism*. Latrobe, Pa.: Archabbey Press, 1951; New York: Meridian paperbound.

Panofsky, Erwin. *Meaning in the Visual Arts*. New York: Anchor, 1955.

Rapoport, Amos. *House, Form and Culture*. Englewood Cliffs, N.J.: Prentice-Hall, 1969.

————. *The Grass Roots of Art*. New York: Wittenborn, 1955.

————. *Icon and Idea?* Cambridge, Mass.: Harvard University Press, 1955.

Read, Sir Herbert. *The Philosophy of Modern Art*. New York: Horizon, 1953.

Rudofsky, Bernard. *Architecture without Architects*. New York: Museum of Modern Art, 1964.

Rosenberg, Harold. *The Tradition of the New*. New York: Horizon, 1957.

Scheidig, Walther. *Crafts of the Weimar Bauhaus*. New York: Reinhold, 1967.

Sempter, Gottfried. *Wissenschaft, Industrie und Kunst*. Mainz, Germany: Florian Kupferberg, 1966.

Singer, Charles (ed.). *A History of Technology* (5 vols.). Oxford, England: Oxford University Press, 1954–58.

Snaith, William. *The Irresponsible Arts*. New York: Atheneum, 1964.

Von Neumann. *Game Theory*. Cambridge, Mass.: M.I.T. Press, 1953.

Wingler, Hans M. *The Bauhaus*. Cambridge, Mass.: M.I.T. Press, 1969.

Youngblood, Gene. *The Expanded Cinema*. New York: Dutton, 1970.

II) THE PRACTICE OF DESIGN AND ITS PHILOSOPHY

Albers, Anni. *On Designing*. New Haven, Conn.: Pellango Press, 1959.

Anderson, Donald M. *Elements of Design*. New York: Holt, Rinehart & Winston, 1961.

Art Directors' Club of New York. *Symbology*. New York: Hastings House, 1960.

———. *Visual Communication: International*. New York: Hastings House, 1961.

Baker, Stephen. *Visual Persuasion*. New York: McGraw-Hill, 1961.

Bayer, Herbert. *Visual Communication, Architecture, Painting*. New York: Reinhold, 1967.

Bill, Max. *Form* (text in German, English, French). Basel, Switzerland: Karl Werner, 1952.

Doxiadis, Constantinos. *Architecture in Transition*. New York: Oxford University Press, 1963; London: Hutchinson, 1965.

———. *Between Dystopia and Utopia*. London: Faber & Faber, 1966.

———. *Ekistics*. New York: Oxford University Press, 1968.

Gropius, Walter. *Scope of Total Architecture*. New York: Harper, 1955.

Itten, Johannes. *The Art of Color*. New York: Reinhold, 1961.

———. *Design and Form*. New York: Reinhold, 1963.

———. *On the Spiritual in Art*. New York: Wittenborn, 1948.

Kandinsky, Wassily. *Point to Line to Plane*. New York: Guggenheim Museum, 1947.

Kepes, Gyorgy. *Language of Vision*. Chicago: Paul Theobald, 1949.

———. *The New Landscape in Art and Science*. Chicago: Paul Theobald, 1956.

———. *Vision-Value Series*. New York: George Braziller, 1966.

Vol. 1. *Education of Vision*.

Vol. 2. *Structure in Art and Science*.

Vol. 3. *The Nature and Art of Motion*.

Vol. 4. *Module Proportion Symmetry Rhythm*.

Vol. 5. *The Man-made Object*.

Vol. 6. *Sign, Image, Symbol*.

———. (ed.). *The Visual Arts Today*. Middletown, Conn.: Wesleyan University Press, 1960.

Kuebler, George. *The Shape of Time*. New York: Schocken, 1967.

Klee, Paul. *Pedagogical Sketch Book*. New York: Praeger, 1953.

———. *The Thinking Eye*. New York: Wittenborn, 1961.

Kranz, Stewart, and Fisher, Robert. *The Design Continuum*. New York: Reinhold, 1966.

Malevich, Kasimir. *The Non-objective World*. Chicago: Paul Theobald, 1959.

Moholy-Nagy, László. *The New Vision* (4th Edition). New York: Wittenborn, 1947.

———. *Telehor*. Bratislava, Czechoslovakia: 1938.

———. *Vision in Motion*. Chicago: Paul Theobald, 1947.

Moholy-Nagy, Sibyl. *Moholy-Nagy: Experiment in Totality*. New York: Harper, 1950.

Mondrian, Piet. *Plastic and Pure Plastic Art.* New York: Wittenborn, 1947.

Mundt, Ernest. *Art, Form & Civilization.* Berkeley, Cal.: University of California Press, 1952.

Nelson, George. *Problems of Design.* New York: Whitney Publications, 1957.

Newton, Norman T. *An Approach to Design.* Boston: Addison-Wesley Press, 1951.

Niece, Robert C. *Art: An Approach.* Dubuque, Iowa: William C. Brown & Co., 1959.

Rand, Paul. *Thoughts on Design.* New York: Wittenborn, 1947.

12) INDUSTRIAL AND PRODUCT DESIGN

Beresford, Evans J. *Form in Engineering Design.* Oxford, England: Clarendon Press, 1954.

Braun-Feldweg, Wilhelm. *Normen und Formen industrieller Produktion.* Ravensburg, Germany: Otto Maier, 1954.

————. *Industrial Design Heute.* Hamburg, Germany: Rowohlt, 1966.

Chase, Herbert. *Handbook on Designing for Quantity Production.* New York: McGraw-Hill, 1950.

The Design Collection: Selected Objects. New York: Museum of Modern Art, 1970.

Design Forecast No. 1 & No. 2 (published by the Aluminum Company of America). Pittsburgh: 1959, 1960.

Doblin, Jay. *One Hundred Great Product Designs.* New York: Reinhold, 1970.

Drexler, Arthur. *Introduction to Twentieth Century Design.* New York: Museum of Modern Art, 1959.

————. *The Package.* New York: Museum of Modern Art, 1959.

Dreyfuss, Henry. *Designing for People.* New York: Simon & Schuster, 1951.

Eksell, Olle. *Design = Ekonomi.* Stockholm: Bonniers, 1964.

Farr, Michael. *Design in British Industry.* Cambridge, England: Cambridge University Press, 1955.

Friedman, William. *Twentieth Century Design: U.S.A.* Buffalo, N.Y.: Albright Art Gallery, 1959.

Functie en Vorm: Industrial Design in the Netherlands. Bussum, Holland: Moussault's Uitgeverij, 1956.

Gestaltende Industrieform in Deutschland. Düsseldorf, Germany: Econ, 1954.

Gloag, John. *Self Training for Industrial Designers.* London: George Allen & Unwin, 1947.

Holland, Laurence B. (ed.). *Who Designs America?* New York: Anchor, 1966.

Jacobson, Egbert. *Basic Color.* Chicago: Paul Theobald, 1948.

Johnson, Philip. *Machine Art.* New York: Museum of Modern Art, 1934.

Lippincott, J. Gordon. *Design for Business.* Chicago: Paul Theobald, 1947.

Loewy, Raymond. *Never Leave Well Enough Alone.* New York: Simon & Schuster, 1950.

Noyes, Eliot F. *Organic Design.* New York: Museum of Modern Art, 1941.

Pevsner, Nikolaus: *An Enquiry into Industrial Art in England.* Cambridge, England: Cambridge University Press, 1937.

————. *Pioneers of Modern Design.* New York: Museum of Modern Art, 1949.

Read, Sir Herbert. *Art in Industry.* New York: Horizon, 1954.

Teague, Walter Dorwin. *Design this Day.* New York: Harcourt, Brace, 1940.

Van Doren, Harold. *Industrial Design* (2nd Edition). New York: McGraw-Hill, 1954.

Wallance, Don. *Shaping America's Products.* New York: Reinhold, 1956.

The following magazines were also consulted:

Architectura Cuba (Cuba)
Arkkitehti-Lehti (Finland)
Aspen (U.S.A.)
China Life (Peking)
Craft Horizons (U.S.A.)
Design (England)
Design & Environment (U.S.A.)
Design Quarterly (U.S.A.)
Designcourse (U.S.A.)
Domus (Italy)
Dot Zero (U.S.A.)
Draken (Sweden)
Environment (U.S.A.)
Form (Sweden)
form (Germany)
Graphis (Switzerland)
IDSA Journal (U.S.A.)

Industrial Design (U.S.A.)
Journal of Creative Behavior (U.S.A.)
Kaiser Aluminum News (U.S.A.)
Kenchiko Bunko (Japan)
Mobilia (Denmark)
Newsweek (U.S.A.)
Stile Industria (Italy)
Sweden NOW (Sweden)
Ulm (Germany)
&/sdo (Helsinki and Stockholm)

ERRATA

p. 48   The last line of the caption should read:
collection. Photo: Roger Conrad.)

p. 67   The caption at the bottom of the page refers to the two pictures on p. 68.

p. 69   The picture should carry the following caption:
Another variation of African TV proposal, designed by Stanhope Adams, Jr., as a student at Purdue University.

p. 317   The titles *The Eternal Present* and *The Beginnings of Architecture* should be listed under the name of the author, Sigfried Giedion.

The title *The Beer Can by the Highway* should be listed under the name of the author, John A. Kouwenhoven.

The titles *The Confessions of a Dilettante* and *The Domesticated Americans* should be listed under the name of the author, Russell Lynes.

The titles *The Brown Decades, The City in History, The Condition of Man, The Conduct of Life, The Culture of Cities, From the Ground Up,* and *Sticks and Stones* should all be listed under the name of the author, Lewis Mumford.

p. 319   Lines 27–29, beginning "All the above and future documents . . . ," should immediately follow R. Buckminster Fuller, *The Ecological Context: Energy and Materials.*

p. 321   The title *The Study of Instinct* should appear under the name of the author, Nicolaas Tinbergen.

p. 331   The title *The Bolivian Diaries of Che Guevara* should be listed under the name of the author, Che Guevara.

p. 335   The title *Gothic Architecture and Scholasticism* should be listed under the name of the author, Erwin Panofsky.

The titles *The Grass Roots of Art* and *Icon and Idea* should appear under the name of the author, Sir Herbert Read.

p. 336   The title *On the Spiritual in Art* should appear under the name of the author, Wassily Kandinsky.